PLANET EARTH
LABORATORY MANUAL

UNC GEOLOGICAL SCIENCES
DR. MEGAN PLENGE

Planet Earth | Laboratory Manual

UNC Geological Sciences
Dr. Megan Plenge

Printed in the United States of America
10 9 8 7 6
ISBN: 978-1-61740-722-2

Van-Griner Learning
Cincinnati, Ohio
www.van-griner.com

President: Dreis Van Landuyt
Project Manager: Maria Walterbusch
Customer Care Lead: Lauren Houseworth

Plenge 722-2 F19
316792-325243
Copyright © 2021

TABLE OF CONTENTS

LABORATORIES

PREFACE

PURPOSE OF THE LAB

The lab section of GEOL 101 is intended to give you hands-on experience with the types of problems, measurements, calculations, and materials geoscientists use. Little to no lecture time will be included in each laboratory section. If you're feeling underprepared for completing the lab activities because of this, please make sure you are spending sufficient time on the pre-lab activities.

PRE-LAB ACTIVITIES

The lab broadly corresponds with the GEOL 101 lectures; however, the content in the lab may be different than what is presented in the lecture. Therefore, pre-lab activities are essential in preparing you for each week's laboratory. Pre-lab activities are provided on the Sakai site and may take the form of one or more readings, videos, or a combination of the two. Please take notes and study pre-lab readings and videos extensively, as this is your best way to prepare for the lab.

PRE-LAB QUIZZES

To hold students accountable for preparing for lab each week, a quiz on the pre-lab activities will be given on Sakai. For full credit, the quiz must be completed prior to the start of your lab section. If you forget to do the quiz prior to your lab section, you have until noon the Friday after your lab section to complete the pre-quiz with a 50% penalty.

Pre-lab quizzes are *open note* and must be completed *individually.* Pre-lab quizzes are timed so that you have 3 minutes per question. Make sure to take notes on the pre-lab activities prior to opening the quiz to help you out!

LAB BOOK

This lab book walks you through the activities you're expected to do each week. Your TA will grade lab book materials for completion, as the main grade you'll be awarded for each lab will be based on the lab report you write. Make sure you ask a lot of questions, confirm answers with your group members and your TA, and understand lab materials each week, as this will help you complete the lab report.

You may collaborate with group members on the Lab Book (completion grade) portion of the labs. You *must* be in class to receive the completion grade; e.g., filling out the lab book at home and then showing it to your TA is not sufficient.

LAB REPORTS

After completing your lab each week, you will write an ***individual*** lab report. Lab reports are due by the start of class 1 week after in-class completion. That should give you time to reflect on the lab book materials, go to office hours for help, etc.

Lab reports are *open note* and must be completed *individually*. Evidence of collusion or collaboration with others will result in a zero on the assignment.

LAB REPORT C–E–R FORM

This part of the lab report should illustrate that you understand why we did the lab, how the data we collected in the lab tells a story, and why the data we collected is informative. Include the following:

- *__The question you are trying to answer:__* You will always be provided a question that the lab is trying to answer, but you're welcome to come up with your own question as well.

- *__A claim:__* *The claim is the conclusion of your experiment:* the answer to the question you're trying to answer! **You will develop your own,** and it should be unique (e.g., no collaboration with teammates). The claim may be in the form of a hypothesis or a prediction. It does not have to encompass everything that you discovered during the lab but should encompass the major aspects.

- **Evidence:** The data you collected during the lab period that makes you think your claim is correct. Different people may include different things here based on their claim.

- **Reasoning:** It may not be clear to the person reading your C–E–R form why you think the evidence you included is sufficient evidence that your claim is correct. Prove it to them in this section. It is often helpful to number the evidence and reasoning sections so that you have specific reasons each for the validity of each piece of evidence. However, some people choose not to number it.

LAB REPORT GRADING

The lab report template can be found on the following page, along with the rubrics for grading. The point break-down is as follows:

1. Lab book completion/attendance: 8 points total
2. C–E–R form: 12 points total (see rubric for point break-down)

POST-LAB QUIZZES

In order to test whether or not you have achieved each lab's learning objectives, the Sakai quizzes that contain pre-lab questions will also have questions covering learning objectives from the previous week. For full credit, the quiz must be completed prior to the start of your lab section. If you forget to do the quiz prior to your lab section, you have until noon the Friday after your lab section to complete the pre-quiz with a 50% penalty.

Post-lab and pre-lab quiz questions will be found in a single quiz you take each week. Post-lab quizzes are *open note* and must be completed *individually*. Quizzes are timed so that you have 3 minutes per question. Make sure to have your lab book available to help you answer questions!

RUBRIC FOR COMPLETION/ATTENDANCE

	1	2	3	4
Score ×2 for total of 10 points *(–2 pts if student is 10 or more minutes late to lab, or off-task during lab (e.g., copying from lab partners, on phone)*	The lab hand-out is less than 50% complete	Lab is ~50–70% complete	Lab is ~70–95% complete	Lab is ~95–100% complete

C–E–R TEMPLATE FOR LAB REPORT

Lab Report Part 1: C–E–R

TABLE P.1	
THE QUESTION YOU'RE TRYING TO ANSWER:	
CLAIM: *ONE-SENTENCE ANSWER TO THE QUESTION BEING INVESTIGATED*	
EVIDENCE: *WHAT EVIDENCE/DATA COULD YOU COLLECT THAT WOULD SUPPORT YOUR CLAIM?*	**REASONING:** *WHY WOULD THIS EVIDENCE SUPPORT YOUR CLAIM? JUSTIFY WHETHER OR NOT THIS EVIDENCE IS SUFFICIENT/VALID.*
1.	1.
2.	2.
3.	3.

C–E–R RUBRIC FOR LAB REPORT

	1	2	3	4
Claim (*–1 pt if the claim does not focus on the most important aspect of the question/ experiment*)	The claim has *major* errors in all 3 points as listed below: 1. Wording or length makes the claim unclear. 2. There is a misalignment with either the question or evidence so that some aspect of the claim may be unsupported. 3. The claim is lacking in specificity or detail, such that the evidence provided is not sufficient to support it.	The claim has *minor* errors in all 3 points as listed below or *major* errors in 2 of the 3: 1. Wording or length makes the claim unclear. 2. There is a misalignment with either the question or evidence so that some aspect of the claim may be unsupported. 3. The claim is lacking in specificity or detail, such that the evidence provided is not sufficient to support it.	The claim has *minor* errors in 2 of the 3 points as listed below or *major* errors in 1 of the 3: 1. Wording or length makes the claim unclear. 2. There is a misalignment with either the question or evidence so that some aspect of the claim may be unsupported. 3. The claim is lacking in specificity or detail, such that the evidence provided is not sufficient to support it.	The claim is well-written, succinct, and aligns logically with both the question and the evidence. The claim focuses on the most important aspect of the question and is specific and detailed enough so that the evidence provided can clearly support it.
Evidence (*–1 pt if the evidence is difficult to follow or understand, e.g. if there is so much content included that it is not clear what the evidence is*)	The evidence has *major* errors in all 3 points as listed below: 1. All or some of the evidence does not align well with the claim or satisfy the claim in its entirety. 2. The evidence lacks in detail; even if it addresses every part of the claim, it does not provide ample evidence. 3. The data included do not include both observation and interpretations.	The evidence has *minor* errors in all 3 points as listed below or *major* errors in 2 of the 3: 1. All or some of the evidence does not align well with the claim or satisfy the claim in its entirety. 2. The evidence lacks in detail; even if it addresses every part of the claim, it does not provide ample evidence. 3. The data included do not include both observation and interpretations.	The evidence has *minor* errors in 2 of the 3 points as listed below or *major* errors in 1 of the 3: 1. All or some of the evidence does not align well with the claim or satisfy the claim in its entirety. 2. The evidence lacks in detail; even if it addresses every part of the claim, it does not provide ample evidence. 3. The data included do not include both observation and interpretations.	The evidence clearly relates to claim, is broad enough to satisfy all parts of the claim, and is detailed enough to provide a convincing case. Both observational and interpreted data are included and presented in an easy-to understand way. Extraneous information that does not directly refer back to the claim or the reasoning should not be included.
Reasoning (*–1 pt if the reasoning is difficult to follow or understand, e.g. if there is so much content included that the grader is unable to evaluate your understanding of the concepts*)	The reasoning has *major* errors in all 3 points as listed below: 1. Background knowledge is not included or is not correct so that the reasoning is incorrect or unclear. 2. The reasoning does not sufficiently explain why or how the evidence supports the claim. 3. The reflection on the validity/quality of the evidence is not sufficiently detailed.	The reasoning has *minor* errors in all 3 points as listed below or *major* errors in 2 of the 3: 1. Background knowledge is not included or is not correct so that the reasoning is incorrect or unclear. 2. The reasoning does not sufficiently explain why or how the evidence supports the claim. 3. The reflection on the validity/quality of the evidence is not sufficiently detailed.	The reasoning has *minor* errors in 2 of the 3 points as listed below or *major* errors in 1 of the 3: 1. Background knowledge is not included or is not correct so that the reasoning is incorrect or unclear. 2. The reasoning does not sufficiently explain why or how the evidence supports the claim. 3. The reflection on the validity/quality of the evidence is not sufficiently detailed.	The reasoning logically connects the evidence and claim using accurate background knowledge. The quality of the evidence is evaluated with specific examples of where evidence is strong and where it needs to be supported further.

GEOSCIENCE RESEARCH

PREFACE

INTRODUCTION

A dialogue you overhear …

Luis: You're taking Intro Geology Lab this semester? Cool!

Kristin: … I don't know about that. I'm not into rocks. But I have to get my PX credit somehow, and I definitely don't want to take chemistry. Geology just seems like the easiest science.

Luis: I don't know if I'd call rock identification easy …

Kristin: Yeah but it's just memorizing stuff, and there's no math so … (shrug). I think all geologists do is look at stuff and describe it, and I can definitely do that.

Luis: I'm not sure that's all geologists do. My neighbor works for the U.S. Geological Survey, and she somehow figured out how old the moon is. I think we're gonna have to do some math.

Kristin: That sounds like a load of B.S. to me. I think they just made stuff up. I mean, how could we *possibly* know something like that—no one was around when the moon formed!

Luis: I don't know exactly what she does, but I DO know that she's doing more than "looking at stuff." I think she can measure something in the moon rocks they've collected to figure out how old they are.

Today, we're going to be doing a brainstorming activity to help you answer this question:

- ⊙ **How do geoscientists conduct research?**

LEARNING OBJECTIVES

Students will be able to …

- ⊙ Compare and contrast hypothesis, prediction, and model.
- ⊙ Use a model to generate a prediction.
- ⊙ Identify types of evidence that can be used to support a hypothesis, and justify why or how this evidence is useful.
- ⊙ Explain why models are useful in conducting geoscience experiments.

BEFORE YOU BEGIN

Look at the syllabus to answer the following questions. The syllabus can be found in the logistics section of Sakai. If you are unclear about anything, the TA will present a powerpoint with information at the beginning of class as well. A PDF of the PowerPoint can also be found on Sakai.

1. What's your TA's name and email address? ________________________________

2. What is the due date for the lab pre-quiz each week? ________________________

3. What is the latest you can do the lab pre-quiz each week? ____________________

4. What do you have to turn in to get credit for completing each lab? ____________
    ____

5. What is the due date for the lab each week? ______________________________

6. What is the latest you can turn in the lab each week? _______________________

7. You are expected to attend your regularly scheduled lab each week. If you have to miss lab one week, can you still get credit for the lab? If so, how? ____________
    ____
    ____

PART 1: HYPOTHESES, MODELS, AND PREDICTIONS ———————

PROCEDURE

1. Complete this Venn diagram contrasting scientific models, hypotheses, and predictions.

   a. Proposed explanation to a scientific question

   b. Use data/prior knowledge

   c. Expected future condition, deduced from a hypothesis

   d. Synthesizes data to test a hypothesis

   e. "The Moon has different phases because the half of the Moon that's facing the Sun is not always facing the Earth."

   f. "If the Earth is between the Moon and the Sun, then the Moon is visible from the Earth as a full moon."

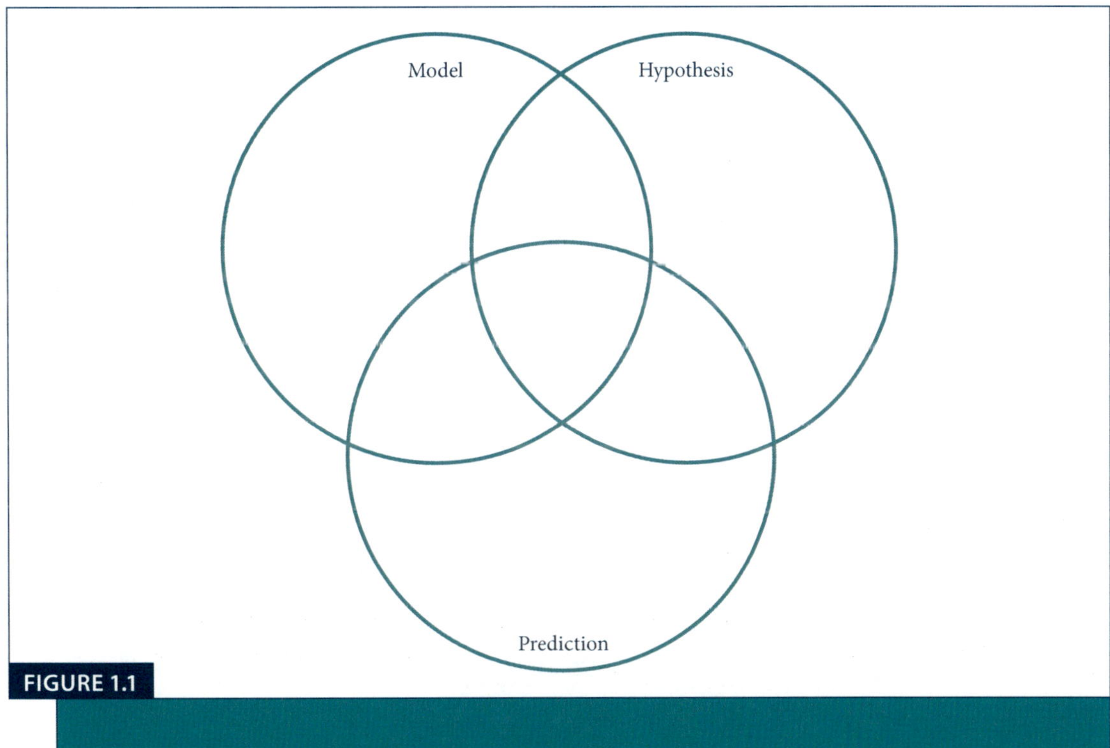

FIGURE 1.1

2. The model below will help you visualize how the Moon should look from the Earth if the hypothesis, "The Moon has different phases because the half of the Moon that's facing the Sun is not always facing the Earth," is true.

    a. Imagine you are standing directly under each of the Moon models on Earth. If you looked straight up, you would see only the half of the Moon that is facing the Earth.

        i. Draw a line dividing each of the interior Moon models in half based on which side is facing the Earth. #7 has been done as an example for you.

        ii. In each of the exterior Moon models, draw what you **should** see on Earth based on looking only at the half of the Moon that is facing the Earth in each position.

3. Watch the video/model shown here to check your work and to label the name of each Moon phase. *http://www.sumanasinc.com/webcontent/animations/content/moonphase.html*

    a. Phase 1: ____

    b. Phase 2: ____

    c. Phase 3: ____

    d. Phase 4: ____

    e. Phase 4: ____

    f. Phase 5: ____

    g. Phase 6: ____

    h. Phase 7: ____

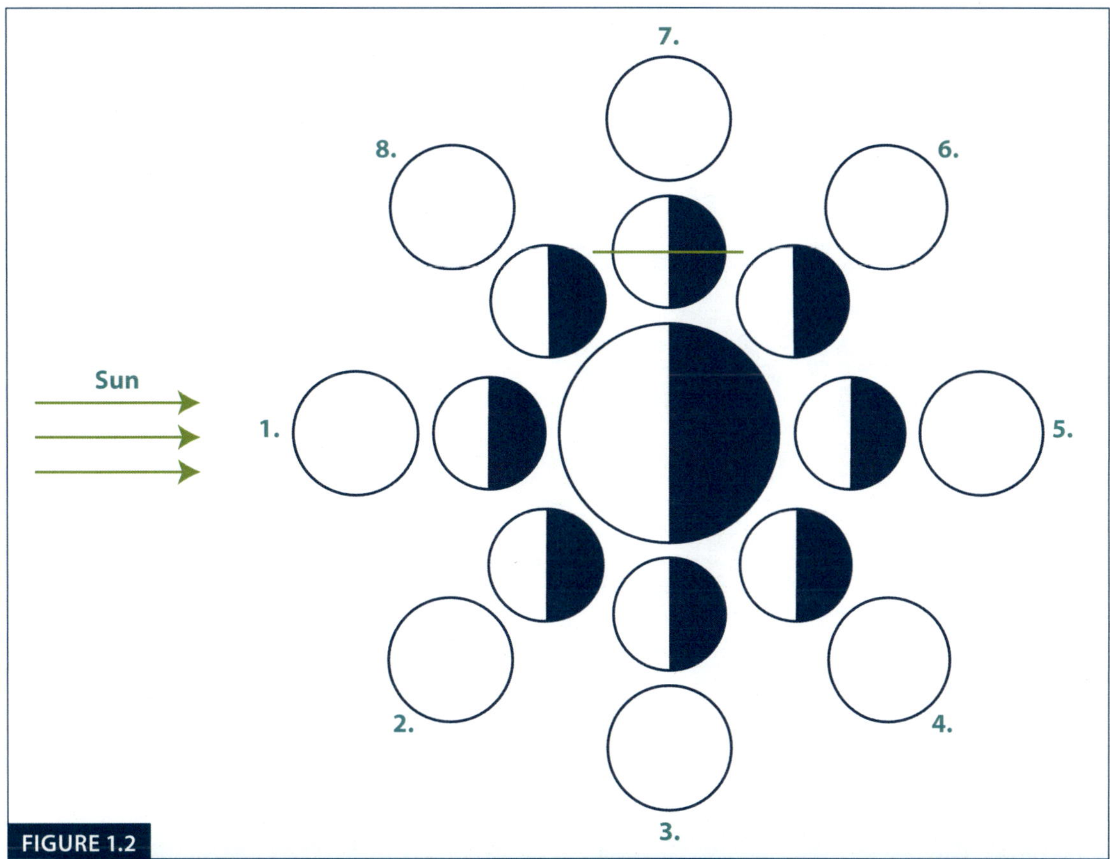

FIGURE 1.2

Visual model of the Sun-Earth-Moon system. The Sun is off to the left of the diagram. The Earth is the large circle in the center. Eight different positions the Moon can be in relative to the Earth and Sun are numbered in a counterclockwise manner, consistent with the direction the Moon revolves around the Earth. The Earth and the 8 interior Moon models have been shaded so that the side that is not facing the Sun appears dark.

4. Brainstorm with your group to answer the following questions:

   a. **Does the model above correspond well with the hypothesis,** "The Moon has different phases because the half of the Moon that's facing the Sun is not always facing the Earth"? How could we improve the visual model?

   b. **Can this model be used to test the prediction,** "If the Earth is between the Moon and the Sun, then the Moon is visible from the Earth as a full moon"? Justify your answer. Would changing the model alter our ability to test this prediction?

c. **Make a prediction** about the time of day one or more of the Moon phases would be visible based on the model.

d. **What evidence could you collect …**

i. that would illustrate that your prediction was correct?

ii. that would illustrate that your prediction was incorrect?

e. A hypothesis by definition is falsifiable. **Could the type of evidence you collected to test your prediction in #7 be used to falsify the hypothesis:** "The Moon has different phases because the half of the Moon that's facing the Sun is not always facing the Earth"? **If yes, explain why. If no, list types of evidence you could collect to falsify the hypothesis.**

PART 2: BRAINSTORMING TO CONNECT QUESTIONS, HYPOTHESES, EVIDENCE, AND REASONING USING THE C–E–R METHOD

PROCEDURE

In lab this semester, we will be outlining what we're doing and why we're using the C–E–R method. This method breaks down each lab into parts:

- *The question you are trying to answer:* You will always be provided a question that the lab is trying to answer, but you're welcome to come up with your own question as well.

- *A claim:* **The claim is the conclusion of your experiment:** the answer to the question you're trying to answer! You will develop your own, and it should be unique (e.g., no collaboration with teammates) for labs this semester. The claim may be in the form of a hypothesis or a prediction. It does not have to encompass everything that you discovered during the lab but should encompass the major aspects.

- *Evidence:* The data you collected during the lab period that makes you think your claim is correct. Different people may include different things here based on their claim.

- *Reasoning:* It may not be clear to the person reading your C–E–R form why you think the evidence you included is sufficient evidence that your claim is correct. Prove it to them in this section. It is often helpful to number the evidence and reasoning sections so that you have specific reasons for the validity of each piece of evidence. However, some people choose not to number it.

1. Work with your group and use your answers from #4 to help you complete the C–E–R form. The question has been filled in for you. You do not have to have 3 distinct lines of evidence/reasoning.

The C–E–R method.

TABLE 1.1

THE QUESTION YOU'RE TRYING TO ANSWER:
Why do we see different phases of the Moon at different times of day?

CLAIM: *ONE-SENTENCE ANSWER TO THE QUESTION BEING INVESTIGATED*

EVIDENCE: *WHAT EVIDENCE/DATA COULD YOU COLLECT THAT WOULD SUPPORT YOUR CLAIM?*	REASONING: *WHY WOULD THIS EVIDENCE SUPPORT YOUR CLAIM? JUSTIFY WHETHER OR NOT THIS EVIDENCE IS SUFFICIENT/VALID.*
1.	1.
2.	2.
3.	3.

2. ***Switch groups.*** Your TA will facilitate you moving to work with different people. With this group, read the question and claim provided below and brainstorm the types of evidence that could support this claim and your reasoning for why this evidence is sufficient/valid.

The C–E–R method.

TABLE 1.2

THE QUESTION YOU'RE TRYING TO ANSWER:
Why does the melting of sea ice change the salinity of the oceans?

CLAIM: *ONE-SENTENCE ANSWER TO THE QUESTION BEING INVESTIGATED*
Sea ice is fresh water because salt doesn't fit into the crystal structure of ice, so when it melts, the salinity of the ocean decreases.

EVICENCE: *WHAT EVIDENCE/DATA COULD YOU COLLECT THAT WOULD SUPPORT YOUR CLAIM?*	REASONING: *WHY WOULD THIS EVIDENCE SUPPORT YOUR CLAIM? JUSTIFY WHETHER OR NOT THIS EVIDENCE IS SUFFICIENT/VALID.*
1.	1.
2.	2.
3.	3.

3. What kind of experiment could you conduct in a laboratory setting to collect the evidence you need to support your claim? Draw or explain the general set-up. You could also use computer models, satellite imagery, maps, and other data available online.

4. What kinds of evidence/data could falsify your claim?

5. ***Switch groups.*** Your TA will facilitate you moving to work with different people. With this group, read the question and claim provided below and brainstorm the types of evidence that could support this claim and your reasoning for why this evidence is sufficient/valid.

The C–E–R method.

TABLE 1.3	
THE QUESTION YOU'RE TRYING TO ANSWER:	
How do large Saharan dust storms affect the global climate?	
CLAIM: ***ONE-SENTENCE ANSWER TO THE QUESTION BEING INVESTIGATED***	
Large Saharan dust storms decrease global temperatures by blocking solar radiation from reaching Earth.	
EVIDENCE: ***WHAT EVIDENCE/DATA COULD YOU COLLECT THAT WOULD SUPPORT YOUR CLAIM?***	**REASONING:** ***WHY WOULD THIS EVIDENCE SUPPORT YOUR CLAIM? JUSTIFY WHETHER OR NOT THIS EVIDENCE IS SUFFICIENT/VALID.***
1.	1.
2.	2.
3.	3.

6. What kind of experiment could you conduct in a laboratory setting to collect the evidence you need to support your claim? Draw or explain the general set-up. You could also use computer models, satellite imagery, maps, and other data available online.

7. What kinds of evidence/data could falsify your claim?

PART 3: GRADING A C–E–R

1. Watch this video: *https://youtu.be/MMbsGHVzXRU*

2. Read the 2 C–E–R forms below, and grade them according to the rubric provided. These C–E–R forms aren't exactly like what you'll be doing in class: You'll be doing a lab and then writing about it, while these people just watched the video and wrote about it.

3. Grade each C–E–R form using the rubric in Appendix 1.

The C–E–R Method Form 1.

TABLE 1.4	
THE QUESTION YOU'RE TRYING TO ANSWER:	
What forms the weird ripples seen in West Bar?	
CLAIM: ***ONE-SENTENCE ANSWER TO THE QUESTION BEING INVESTIGATED***	
The surface features in West Bar were formed by a flood of fast-moving water because ripples that size and shape are formed by fluids moving sediments or rocks.	
EVIDENCE: ***WHAT EVIDENCE/DATA COULD YOU COLLECT THAT WOULD SUPPORT YOUR CLAIM?***	**REASONING:** ***WHY WOULD THIS EVIDENCE SUPPORT YOUR CLAIM? JUSTIFY WHETHER OR NOT THIS EVIDENCE IS SUFFICIENT/VALID.***
1. The surface features in West Bar look like a large-scale version of active current ripples that wind forms in sand. 2. Up to 5-foot boulders can be found in the ripples in West Bar. Ripples are 20 feet high and up to 100 yards from ridge to ridge. 3. The ripples cover a large area and are found in several states.	1. The fact that the same patterns are seen in both settings indicate that their method of formation is similar. This indicates that a fluid moving sediments could create this shape. 2. The scale of the sediments that were moved and the size of the ripple features themselves indicate that there had to be a lot of fast-moving water to create these features. The velocity of water is directly related to the amount and size of the sediment it can carry. 3. This supports the idea that there was a "flood" or a lot of water. If these features formed by fast-moving water and there are hundreds of miles between the sites where they are found, there must have been A LOT of water. I don't think the evidence included here is sufficient to make this claim. For example, they give us the measurements of the height and the space between ripple crests, and the size of some of the boulders, but they don't actually talk about the velocity of water that would be required to create that shape. We would need to do an experiment to see if there's a ratio that indicates water has deposited something. They also don't talk about how they know when this formed or where the lake would be or anything.

The C–E–R Method Form 2.

TABLE 1.5

THE QUESTION YOU'RE TRYING TO ANSWER:
What forms the weird ripples seen in West Bar?

CLAIM: *ONE-SENTENCE ANSWER TO THE QUESTION BEING INVESTIGATED*
A flood of water created ripples in the land.

EVIDENCE: *WHAT EVIDENCE/DATA COULD YOU COLLECT THAT WOULD SUPPORT YOUR CLAIM?*	REASONING: *WHY WOULD THIS EVIDENCE SUPPORT YOUR CLAIM? JUSTIFY WHETHER OR NOT THIS EVIDENCE IS SUFFICIENT/VALID.*
1. The shape of the land 2. The size of the rocks 3. The number of ripples	1. The ripples in the land look like sand ripples. 2. Large rocks can only be moved by water. 3. There are a lot of ripples, so there was a lot of water. This argument sounds pretty convincing, but I've never heard of Lake Missoula. Is that real?

TABLE 1.6

RUBRIC SCORING	C–E–R 1			C–E–R 2		
	INDIVIDUAL	GROUP	CLASS	INDIVIDUAL	GROUP	CLASS
Claim						
Evidence						
Reasoning						

APPENDIX 1: C–E–R RUBRIC

	1	2	3	4
Claim *(–1 pt if the claim does not focus on the most important aspect of the question/experiment)*	The claim has *major* errors in all 3 points as listed below: 1. Wording or length makes the claim unclear. 2. There is a misalignment with either the question or evidence so that some aspect of the claim may be unsupported. 3. The claim is lacking in specificity or detail, such that the evidence provided is not sufficient to support it.	The claim has *minor* errors in all 3 points as listed below or *major* errors in 2 of the 3: 1. Wording or length makes the claim unclear. 2. There is a misalignment with either the question or evidence so that some aspect of the claim may be unsupported. 3. The claim is lacking in specificity or detail, such that the evidence provided is not sufficient to support it.	The claim has *minor* errors in 2 of the 3 points as listed below or *major* errors in 1 of the 3: 1. Wording or length makes the claim unclear. 2. There is a misalignment with either the question or evidence so that some aspect of the claim may be unsupported. 3. The claim is lacking in specificity or detail, such that the evidence provided is not sufficient to support it.	The claim is well-written, succinct, and aligns logically with both the question and the evidence. The claim focuses on the most important aspect of the question and is specific and detailed enough so that the evidence provided can clearly support it.
Evidence *(–1 pt if the evidence is difficult to follow or understand, e.g. if there is so much content included that it is not clear what the evidence is)*	The evidence has *major* errors in all 3 points as listed below: 1. All or some of the evidence does not align well with the claim or satisfy the claim in its entirety. 2. The evidence lacks in detail; even if it addresses every part of the claim, it does not provide ample evidence. 3. The data included do not include both observation and interpretations.	The evidence has *minor* errors in all 3 points as listed below or *major* errors in 2 of the 3: 1. All or some of the evidence does not align well with the claim or satisfy the claim in its entirety. 2. The evidence lacks in detail; even if it addresses every part of the claim, it does not provide ample evidence. 3. The data included do not include both observation and interpretations.	The evidence has *minor* errors in 2 of the 3 points as listed below or *major* errors in 1 of the 3: 1. All or some of the evidence does not align well with the claim or satisfy the claim in its entirety. 2. The evidence lacks in detail; even if it addresses every part of the claim, it does not provide ample evidence. 3. The data included do not include both observation and interpretations.	The evidence clearly relates to claim, is broad enough to satisfy all parts of the claim, and is detailed enough to provide a convincing case. Both observational and interpreted data are included and presented in an easy-to understand way. Extraneous information that does not directly refer back to the claim or the reasoning should not be included.
Reasoning *(–1 pt if the reasoning is difficult to follow or understand, e.g. if there is so much content included that the grader is unable to evaluate your understanding of the concepts)*	The reasoning has *major* errors in all 3 points as listed below: 1. Background knowledge is not included or is not correct so that the reasoning is incorrect or unclear. 2. The reasoning does not sufficiently explain why or how the evidence supports the claim. 3. The reflection on the validity/quality of the evidence is not sufficiently detailed.	The reasoning has *minor* errors in all 3 points as listed below or *major* errors in 2 of the 3: 1. Background knowledge is not included or is not correct so that the reasoning is incorrect or unclear. 2. The reasoning does not sufficiently explain why or how the evidence supports the claim. 3. The reflection on the validity/quality of the evidence is not sufficiently detailed.	The reasoning has *minor* errors in 2 of the 3 points as listed below or *major* errors in 1 of the 3: 1. Background knowledge is not included or is not correct so that the reasoning is incorrect or unclear. 2. The reasoning does not sufficiently explain why or how the evidence supports the claim. 3. The reflection on the validity/quality of the evidence is not sufficiently detailed.	The reasoning logically connects the evidence and claim using accurate background knowledge. The quality of the evidence is evaluated with specific examples of where evidence is strong and where it needs to be supported further.

2

MINERALS AND IGNEOUS ROCKS

Name: _____________________________ Section: __________ Date: _________

Some sections are adapted from NC State University's MEA 110 Mineral Lab by David McConnell, Katherine Ryker, April Grissom (Jones) and Doug Czajka, Geoscience Learning Process Research group, Department of MEAS, NC State. Mineral Flow Charts from NCSU.

PREFACE

INTRODUCTION

A dialogue you overhear …

Kristin: Ugh, I knew we were going to have to identify minerals. This is so pointless.

Luis: But isn't it kind of cool to categorize things?

Kristin: You would think that.

De'Nesha: Well, if you know how to identify minerals, you can learn about the geologic history of an area…how a place formed, if valuable minerals might be found there…. It *is* pretty cool actually.

Kristin: Where did that come from?

De'Nesha: I don't know. Maybe it's because of this ring my mom gave me. It's made of citrine (yellow) quartz and it's really beautiful. Someone asked me the other day if it was a yellow diamond—those are

FIGURE 2.1

Citrine quartz. Currently $10–$30/carat. Rama
[CC BY-SA 3.0 fr (https://creativecommons.org/licenses/by-sa/3.0/fr/deed.en)] https://commons.wikimedia.org/wiki/File:Citrine_quartz-AMGL_79477-P5030194-white.jpg

FIGURE 2.2

Yellow diamond. Currently $7,000/carat. Rob Lavinsky, iRocks.com – CC-BY-SA-3.0 [CC BY-SA 3.0 (https://creativecommons.org/licenses/by-sa/3.0)] https://commons.wikimedia.org/wiki/File:Diamond-dtn4a.jpg

like suuuuuper expensive. It got me thinking about how people know the difference between different minerals. I mean, how *do* people tell the distinguish between different minerals?

Today, we're giving you the supplies you need to help resolve De'Nesha's question:

- **How can you distinguish between different minerals?**

LEARNING OBJECTIVES

Students will be able to …

- Describe how to classify minerals using physical properties, including hardness, cleavage, color, luster, crystal habit, and reaction with acid.
- Determine the cleavage of an unknown mineral sample.
- Measure the hardness of an unknown mineral sample.
- Use physical properties to identify the following minerals: pyroxene, amphibole, plagioclase feldspar, orthoclase feldspar, quartz, biotite mica, calcite, halite, and pyrite.

MATERIALS

- Hand lens
- Rocks and minerals
- Mineral ID kit
- Balances
- Graduated cylinder

BEFORE YOU BEGIN

BACKGROUND

Please read this before lab.

Today you'll be describing and identifying minerals by using descriptive flow charts that help you differentiate between minerals based on diagnostic physical properties. Use the information below, the appendices at the back of this lab, and the information you learned in the pre-lab videos to help you.

A FEW NOTES ABOUT MINERAL IDENTIFICATION

- **Hardness** is how easily a mineral can be scratched. Objects of known hardness can be used to scratch a mineral surface. If it can be scratched, it's softer than that mineral/object; if not, it is harder. Minerals have been ranked in terms of relative hardness in the Mohs hardness scale. The Mohs hardness scale can be found in an appendix to this lab.

- **Cleavage** is the tendency of a mineral to break along specific planes or in certain directions. Minerals with cleavage have weak bonds in specific places of their crystal structure, so that they break preferentially along those bonds. These breaks can often be identified because those planes will have high luster, and parallel breakage or stair-stepping is often seen. It is easy to confuse cleavage surfaces with crystal faces; typically, minerals with a hardness of 7+ do not cleave, and so flat planes are related to crystal habit. **Fracture** describes minerals that randomly break and do not exhibit any breakage patterns.

- **Crystal habit** is the shape of mineral crystals based on the molecular structure of the mineral. If minerals have time and room to grow, they will always display the same crystal habit. If they are squished against other minerals when growing, it may be more difficult to observe. Note that crystal habit refers to the shape of the mineral, not a breakage pattern.

- **Color** is an unreliable way of identifying minerals, as often minerals are many different colors. For example, quartz can be yellow (citrine), purple (amethyst), grey (smoky), white (milky), pink (rose), clear, or many other colors. Characteristics such as density, crystal structure, fracture pattern, and hardness are much more useful in distinguishing quartz from other minerals, though there are some minerals that have characteristic colors.

- **Luster** describes how light reflects off a mineral's surface. Most minerals will fall into these categories:
  - **Non-metallic** (glassy or dull … believe it or not, glassy can describe something that is not transparent)
  - **Metallic** (the mineral looks metallic)

- **Density/specific gravity** is the mass per unit volume of a mineral.

- **Effervescence** is a fizzing reaction that occurs when a calcium carbonate mineral contacts acid.

- **Magnetism** is a property observed when a mineral can attract a common magnet.

- **Streak** is the color of a powdered mineral. Usually streak is tested by rubbing a mineral across a porcelain plate.

PART 1: WORKING WITH YOUR GROUPS TO TEST MINERAL PROPERTIES

PROCEDURE

1. **Hardness:** Use minerals 1–3 from the mineral box.

   a. Place the piece of glass flat on the lab table. Drag a mineral across the piece of glass. Rub off any powder that formed. If no mark remains, do the next test. If a mark remains, the mineral is harder than the glass, and you can write ">5.5" in the correct spot on Table 1.

   b. If the mineral did not scratch the glass, the hardness is <5.5. Drag the mineral across the copper penny. Rub off any powder that formed. If no mark remains, do the next test. If the mark remains, the mineral is harder than the penny, and you can write "3-5.5" for the possible hardness range in the correct spot on Table 1.

   c. If the mineral did not scratch the penny, the hardness is <3. Try to scratch the mineral with your fingernail. If you can, the hardness is <2.5. If not, the hardness is 2.5–3. Fill in the appropriate spot on Table 2.1.

   d. Repeat this for all 3 samples.

2. **Cleavage:** Use minerals 1–3 from the mineral box.

   a. Inspect the mineral to look for roughly parallel flat sides, or roughly parallel stair-stepped sides. The picture on the right illustrates the stair-stepped sides; the top and bottom of the mineral shown represent one cleavage plane. Each set of parallel sides = 1 cleavage plane.

FIGURE 2.3

Calcite. Photo by Megan Plenge

   b. If it's difficult to tell if you're looking at a cleavage plane because there's a lot of fracturing, hold the mineral in the light and move it back and forth to look for a glossy luster. If you're really looking at cleavage planes, you will see a glossy luster. If there are fracture planes, no luster will be visible. *Note:* Sometimes crystal faces also exhibit high luster, so just seeing flat, glossy sides does not mean it is a cleavage plane

   c. To determine if you're looking at crystal habit or cleavage, drag a steel nail across the mineral. If it scratches it, the hardness is <6.5 and it is likely cleavage (but not always). If the mineral is unscratched, the hardness is >6.5 and you're likely looking at crystal habit. (*Note:* the best way to truly figure this out is to hit it with a hammer. But we're not doing that in class).

    d. Based on your answers for a–c, fill out the correct spot in Table 2.1 to indicate if the mineral shows cleavage or fracture. If cleavage, estimate the number of planes.

    e. Repeat this for all 3 samples.

3. **Other identifying characteristics:** Use minerals 1–3 from the mineral box.

    a. Record the color and luster of samples 1–3. For luster, use metallic, dull, or glassy.

    b. Make observations of other distinguishing properties. You could include the following:

        i. Evidence of crystal habit

        ii. Reaction with acid

        iii. Interesting streak color

4. Use your observations and the flow charts in the appendices to identify minerals 1-3.

5. Check with your TA to make sure you're correct!

DATA COLLECTION

Part 1: Identifying minerals.

TABLE 2.1

SAMPLE	MINERAL HARDNESS (RANGE)	CLEAVAGE/ FRACTURE # OF PLANES	COLOR + LUSTER	OTHER IDENTIFYING CHARACTERISTICS	MINERAL ID
1					
2					
3					

PART 2: BECOMING MINERAL EXPERTS

PROCEDURE

Each member of your group will be assigned 3 or 4 minerals. You will complete the same diagnostic tests you performed in Part 1 to identify your minerals. After you complete your tests and ID the minerals, you will meet with others in your class assigned the same minerals to verify your answers. Then, you'll go back to your original groups to share what you learned.

Mineral assignments:

- If you are in a group of 4:
  - **Person A:** Samples 4, 5, 6
  - **Person B:** Samples 7, 8, 9
  - **Person C:** Samples 10, 11, 12
  - **Person D:** Samples 13, 14, 15

- If you are in a group of 3:
  - **Person A:** Samples 4, 5, 6, 7
  - **Person B:** Samples 8, 9, 10, 11
  - **Person C:** Samples 12, 13, 14, 15

1. Work independently to fill out the rows on the table for your assigned minerals.

2. Once you have identified your minerals, the TA will reorganize the room so that everyone who looks at the same minerals will be together (e.g., a Person A group, a Person B group, etc.).

3. Bring your samples to the expert group. Compare samples, discuss the results of your identification, and resolve any differences.

4. Return to your original groups. Teach your other group members about your minerals and fill out the ID table with the help of your teammates.

DATA COLLECTION

Part 2: Identifying minerals.

| TABLE 2.2 | | | | | |
SAMPLE	MINERAL HARDNESS (RANGE)	CLEAVAGE/ FRACTURE # OF PLANES	COLOR + LUSTER	OTHER IDENTIFYING CHARACTERISTICS	MINERAL ID
4					
5					
6					
7					
8					
9					
10					
11					
12					
13					
14					
15					

APPENDIX 1: MINERAL IDENTIFICATION KEYS

1. Does your mineral have a metallic or non-metallic luster?

    a. If metallic, go to mineral *key #1.*

    b. If *not* metallic, look at the color.

        i. If dark (brown, black, or green), go to *key #2.*

        ii. If light or colorless, go to *key #3.*

2. Once you're at the right key,

    a. Check hardness.

    b. Check cleavage/fracture.

    c. Compare other properties.

    d. Make ID.

KEY #1: METALLIC LUSTER

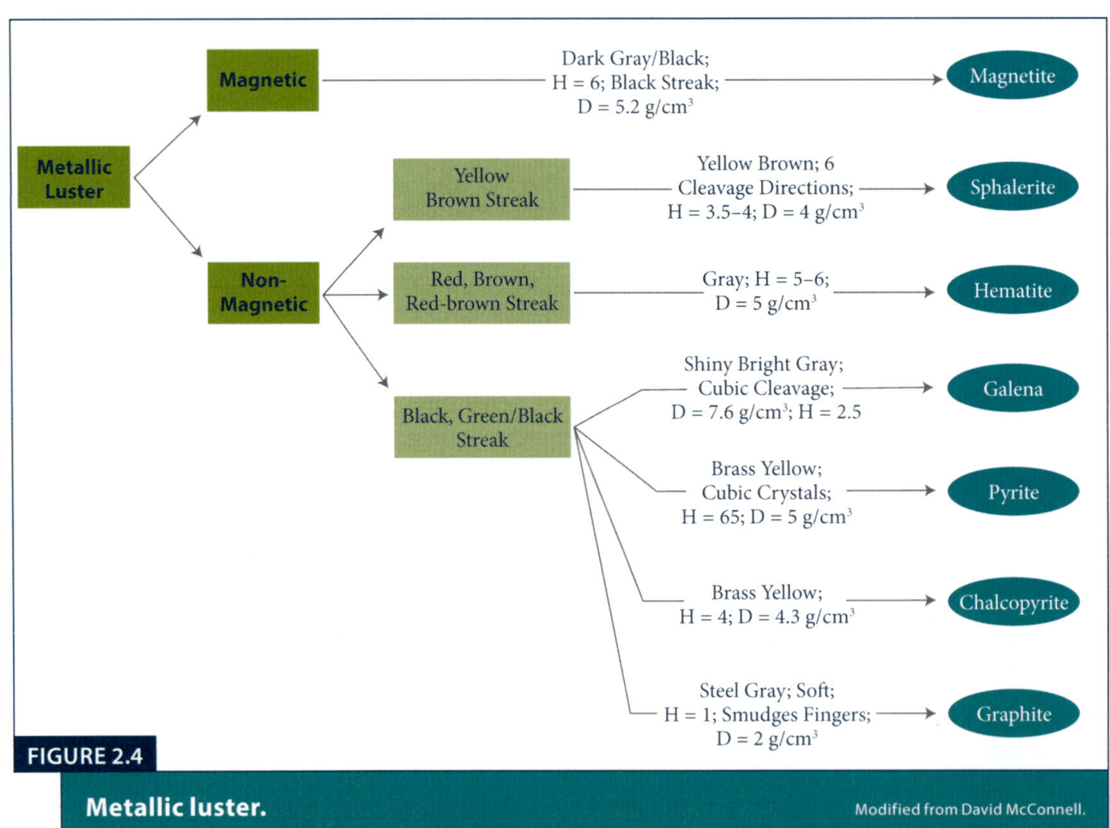

FIGURE 2.4

Metallic luster. Modified from David McConnell.

KEY #2: NON-METALLIC LUSTER, DARK COLOR

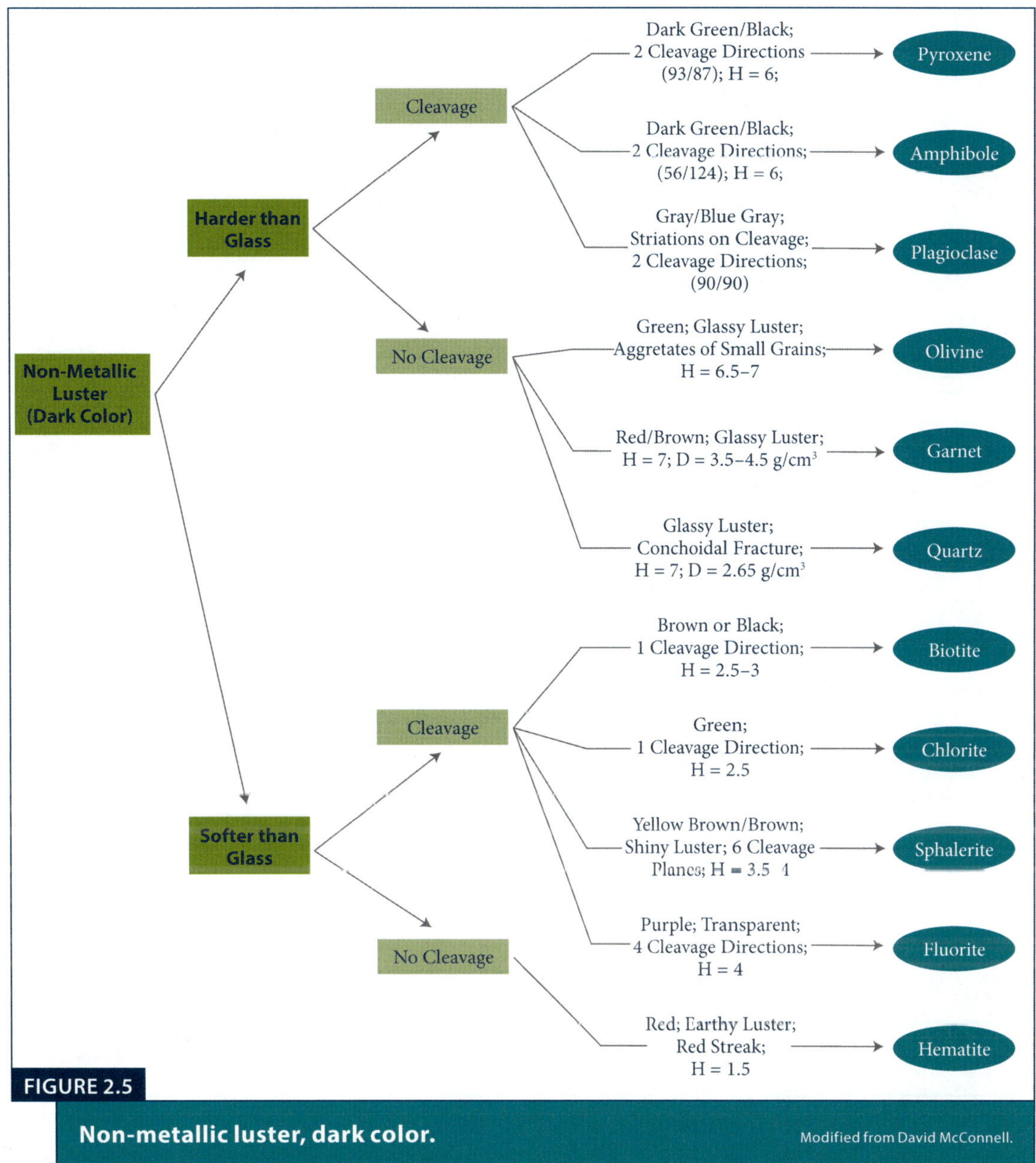

FIGURE 2.5

Non-metallic luster, dark color.   Modified from David McConnell.

KEY #3: NON-METALLIC LUSTER, LIGHT COLOR

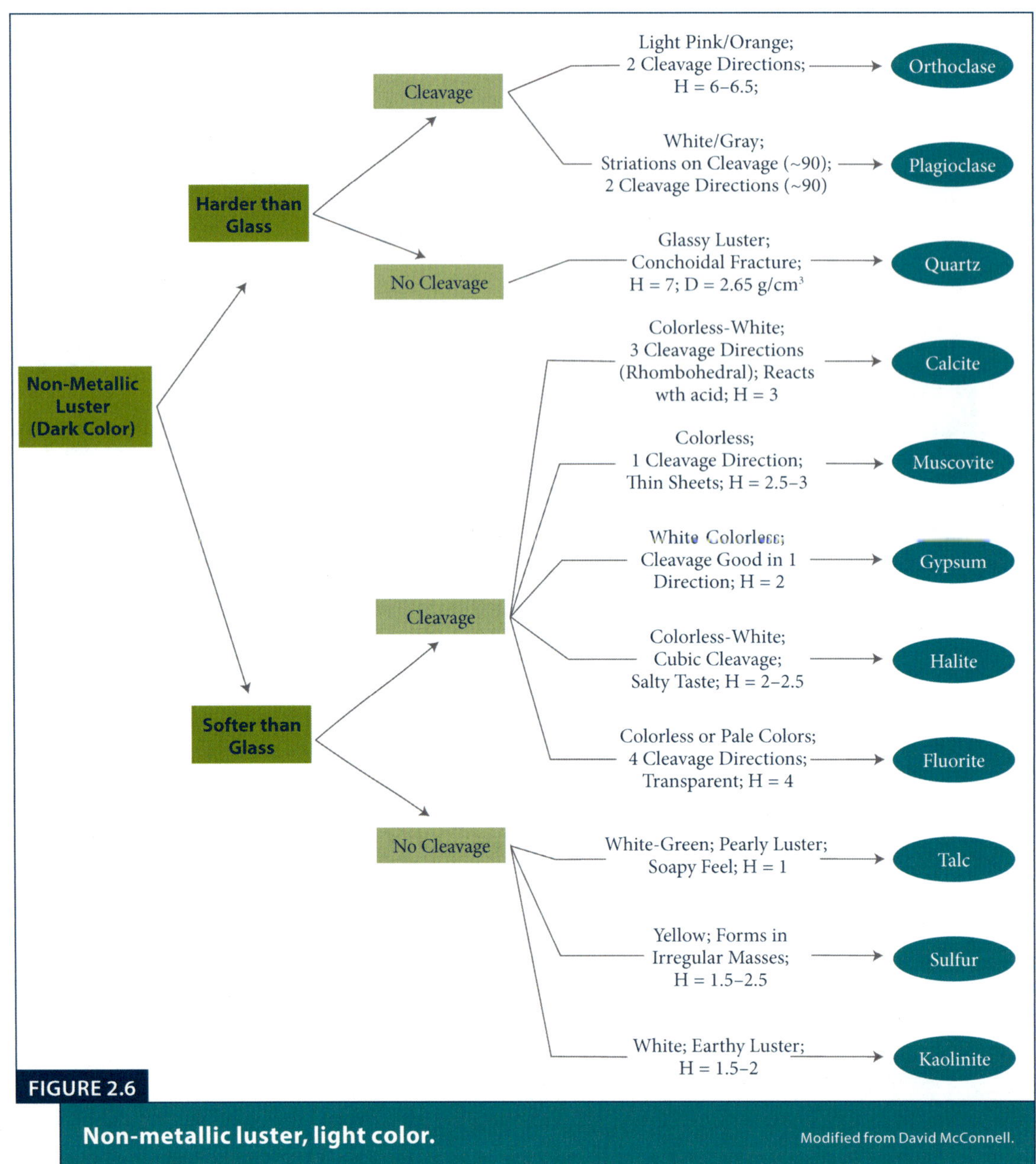

FIGURE 2.6

Non-metallic luster, light color. Modified from David McConnell.

APPENDIX 2: MOHS HARDNESS SCALE

HARDNESS	MINERAL	HOUSEHOLD ITEM
10	Diamond	
9	Corundum	
8	Topaz	
7	Quartz	
6.5		Steel File, Streak Plate
6	Orthoclase Feldspar	
5.5		Glass
5	Apatite	
4.5		Paper Clip
4	Fluorite	
3.5		Copper Penny (pre-1982)
3	Calcite	
2.5		Fingernail
2	Gypsum	
1	Talc	

APPENDIX 3: EXAMPLES OF MINERAL CLEAVAGE

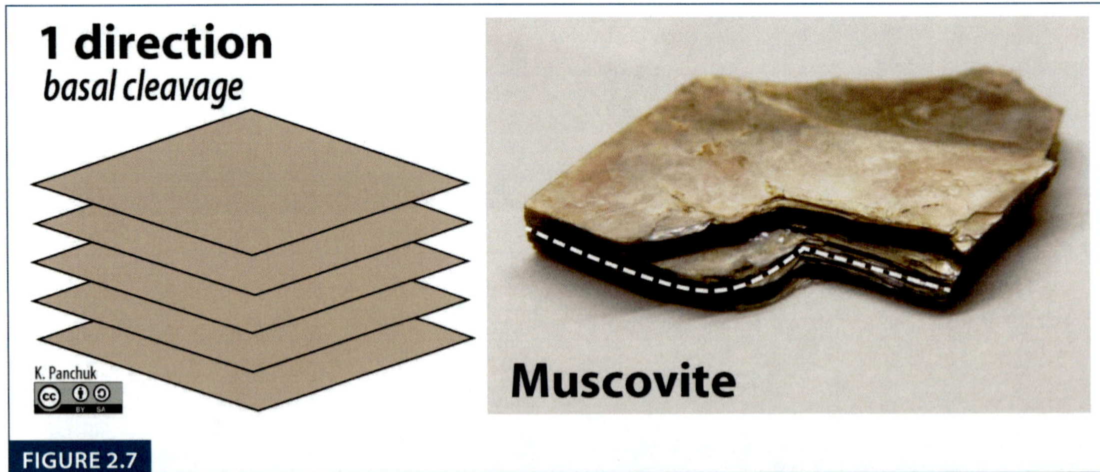

FIGURE 2.7

One direction of cleavage (basal cleavage). Left: Schematic of basal cleavage. Right: Muscovite showing basal cleavage. The white dashed line marks the edge of the cleavage plane.
Source: Karla Panchuk (2018) CC BY-SA 4.0. Cleavage diagram modified after M.C. Rygel (2010) CC BY-SA 3.0

FIGURE 2.8

Two directions of cleavage. Top: Two directions at 90° in pyroxene. Bottom: two directions not at 90° in plagioclase feldspar. Edges of cleavage planes marked with dashed lines.
Source: Karla Panchuk (2018) CC BY-SA 4.0. Cleavage diagrams modified after M.C. Rygel (2010) CC BY-SA 3.0

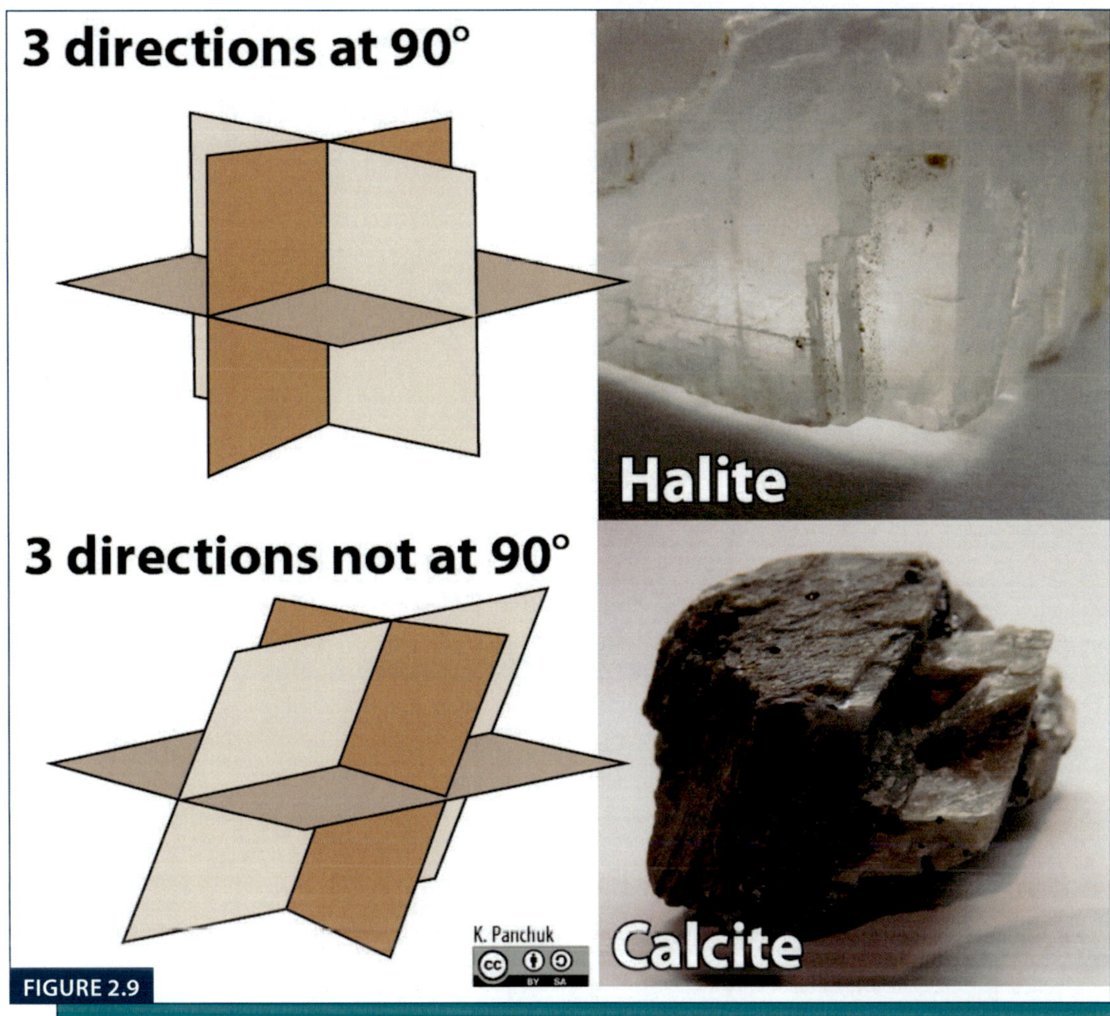

FIGURE 2.9

Three directions of cleavage. Top: Three directions at 90° in halite. Bottom: Three directions not at 90° in calcite. Source: Karla Panchuk (2018) CC BY-SA 4.0. Cleavage diagrams modified after M.C. Rygel (2010) CC BY-SA 3.0

APPENDIX 4: CHART OF BASIC MINERAL CHARACTERISTICS

MINERAL NAME	CLEAVAGE?	AVERAGE DENSITY	HARDNESS	DESCRIPTION (COLOR, LUSTER, OTHER)	IGN.	SED.	META.
Amphibole	2 planes, 56 & 124°	3.3 g/cc	5.5-6	Dark color, needlelike crystals, glassy	X		X
Biotite mica	1 plane	3.1 g/cc	2.5-3	Dark color, glassy	X		X
Calcite	3 planes, not 90°	2.7 g/cc	3	Glassy, reacts with HCl acid		X	X
Chlorite	1 plane	3.0 g/cc	2-2.5	Greenish color			X
Fluorite	4 planes	3.13 g/cc	4	Glassy, many colors	X		X
Galena	3 planes, 90°	7.6 g/cc	2.5-3	Cubic crystals, grey, metallic	X	X	X
Garnet	None, conchoidal fracture	3.7 g/cc	7-7.5	Dodecahedral crystal habit, glassy		X	X
Gypsum	1 plane	2.9 g/cc	2	Glassy or dull white,		X	
Halite	3 planes, 90°	2.1 g/cc	2.5	Salty, glassy		X	
Hematite	None	5–5.3 g/cc	5-6.5	Metallic or dull, rust red streak	X	X	X
Kaolinite	None	2.65 g/cc	2-2.5	Dull, very soft	X	X	X
Magnetite	None	5.17 g/cc	5.5-6.5	Metallic, magnetic	X	X	X
Muscovite mica	1 plane	2.8 g/cc	2.5-3	Light color, vitreous	X	X	X
Olivine	None, uneven fracture	3.3 g/cc	7	Green, glassy, aggregate of small grains	X		X
Orthoclase feldspar	2 planes, 90°	2.5 g/cc	6	Pink or white, glassy along cleavage planes	X	X	X
Plagioclase feldspar	2 planes, 90°	2.7 g/cc	6	White or gray, glassy along cleavage planes	X	X	X
Pyrite	None, conchoidal fracture	5 g/cc	6-6.5	Cubic crystals, grey streak, metallic	X	X	X
Pyroxene	2 planes, ~90°	3.4 g/cc	5.5-6	Dark color, glassy	X		
Quartz	None, conchoidal fracture	2.65 g/cc	7	Hexagonal crystal habit, glassy	X	X	X
Talc	1 plane	2.75 g/cc	1	Pearly, soapy feeling			X

3

IGNEOUS ROCKS AND METAMORPHIC ROCKS

Name: ________________________   Section: __________   Date: _________

Some sections adapted from NC State University's MEA 110 Minerals lab. Mineral flow charts from NCSU.

PREFACE

INTRODUCTION

A dialogue you overhear …

Luis: Here we are in geology lab again!

Kristin: For more rocks. Oh joy.

Luis: Admit it. Last week was more fun than you thought it would be.

Kristin: Nope. I didn't like it. I just didn't feel comfortable—I mean, how are you supposed to know if you're right?

De'Nesha: I'm cool with it. The pre-lab videos made it seem kind of like a puz-zle—you look for clues in the rock to figure out how it was made. And the TA is super helpful. I want her to look at this rock I brought. My little niece gave me—it's got stripes and big crystals (Figure 3.1). I think they're garnets!

Luis: Oooh, lemme see! … Yeah, those do kind of look like the garnet crystals we saw last week. I think it's metamorphic!

Kristin: Do all metamorphic rocks have gemstones in them?

FIGURE 3.1

Garnet gneiss. Public Domain (https://commons.wikimedia.org/wiki/File:Garnet_porphyroblast.JPG#file)

De'Nesha: Ok! So now you're interested! I don't know if all gemstones come from metamorphic rocks. I think the rock type is based on how the rock is made. But maybe all gemstones are made the same way?

Luis: I don't know, but I'm definitely interested in figuring out how a rock is made and what it's made of by observing it!

Today, we're giving you the supplies you need to help Luis figure out:

⊙ **How can we use a rock's color, texture, and other physical characteristics to figure out what a rock is composed of and how it formed?**

LEARNING OBJECTIVES

Students will be able to …

⊙ Explain how to measure the density of a mineral, and identify how density and mineral composition are related to felsic and mafic igneous rocks.

⊙ Categorize igneous rock textures as intrusive, extrusive, or porphyritic and igneous rock mineral compositions as mafic, intermediate, or felsic.

⊙ Categorize metamorphic rock textures as non-foliated or foliated, and identify the metamorphic grade by observing rock texture.

⊙ Interpret rock forming processes (e.g., igneous or metamorphic), and identify igneous and metamorphic rocks given a table of textural and mineralogical characteristics.

MATERIALS

⊙ Hand lens
⊙ Rocks and minerals
⊙ Mineral ID kit
⊙ Balances
⊙ Graduated cylinder

BEFORE YOU BEGIN

Answer the following questions about igneous rocks. Use your pre-lab notes and work together.

1. Explain how the formation of intrusive rocks differs from the formation of extrusive rocks.

2. What characteristics can you use to distinguish intrusive and extrusive igneous rocks?

3. What makes a rock "felsic," "intermediate," or "mafic"? *Be specific (e.g. don't say "one is denser than the other—say which is denser than the other, and name more than one characteristic).*

4. What physical characteristics can you use to distinguish felsic, intermediate, and mafic igneous rocks? *Be specific!*

Answer the following questions about metamorphic rocks. Use your pre-lab and work together.

5. What is the difference in a foliated and non-foliated metamorphic rock?

6. List these metamorphic rocks in order from lowest to highest metamorphic grade:
    ◎ Schist, Gneiss, Phyllite, Slate

7. Which would be more likely to show foliation—a sandstone that metamorphoses into quartzite or a shale that metamorphoses into phyllite? Why?

8. Other than foliation, what characteristics can you look for in a rock to determine whether or not it's metamorphic (e.g., what other visible changes occur during metamorphism)?

PART 1: IGNEOUS AND METAMORPHIC ROCK KIT

The Part 1 rock kit contains two igneous and two metamorphic rocks. You will use what you learned in the pre-lab videos and Appendix 1 of this hand-out to sort these rocks into their respective categories.

PROCEDURE

1. Fill out Table 3.1.

    a. Describe each rock's texture and color.

    b. Identify the rock type: metamorphic (3), igneous (3).

    c. Give a 3–5 word justification of your selection.

DATA COLLECTION

Distinguishing between metamorphic and igneous rocks.

TABLE 3.1			
SAMPLE	**ROCK TEXTURE AND COLOR**	**IGNEOUS OR METAMORPHIC?**	**JUSTIFICATION (2–5 WORDS)**
1A			
1B			
1C			
1D			

1. Which of the following best describes the two igneous rocks? **(Circle one.)**

    a. They are both volcanic/extrusive.

    b. One is volcanic/extrusive and one is plutonic/intrusive.

    c. They are both plutonic/intrusive.

2. List the MAFIC igneous rock sample(s) by sample number: ______________

3. Which of the following best describes the two metamorphic rocks? **(Circle one.)**

    a. They are both foliated   **c.** One is foliated

    b. Neither are foliated

4. List the metamorphic rock(s) that have large (visible) crystals by sample number:

    ____

5. Choose one of the igneous rocks and describe the environment/conditions in which it formed.

    ◉ Sample chosen: ____

    ◉ Description:

6. Choose one of the metamorphic rocks and describe the environment/conditions in which it formed.

    ◉ Sample chosen: ____

    ◉ Description:

PART 2: CALCULATING DENSITY OF ROCKS AND MINERALS

This tray contains 2 minerals and 2 rocks. You'll be determining the density/specific gravity of each sample using this method. Record your data in Table 3.2.

PROCEDURE

1. Turn on the scale if it is not on already. Place your rock on the scale and write down the mass in grams on the chart below.

2. Pour water into your beaker/graduated cylinder and write down the volume of water in mL. If your beaker/graduated cylinder already has water in it, just write down the volume in mL.

3. Place your rock into the water and measure the new volume that is recorded.

4. The volume of the rock is the volume of the water + rock minus the volume of just the water. Record this number in mL.

5. You can leave that rock at the bottom of the beaker. Record new starting volume, add next sample, and measure again.

6. To calculate density, divide mass by volume.

DATA COLLECTION

TABLE 3.2

SAMPLE	MASS, g	VOLUME WATER, mL	VOLUME WATER + SAMPLE, mL	VOLUME SAMPLE, mL	DENSITY SAMPLE, g/mL
2A					
2B					
2C					
2D					

1. Identification and correlation of rock and mineral samples from Part 2.

    a. Identify mineral sample 2A by name using charts from last week's lab:

    ____

    b. Which rock sample likely contains a lot of mineral 2A? **(Circle one.)**

        i. 2C   **ii.** 2D

    c. Identify mineral sample 2B by name using charts from last week's lab:

    ____

    d. Which rock sample likely contains a lot of mineral 2B? **(Circle one.)**

        i. 2C   **ii.** 2D

2. Use the igneous rock identification chart in the Appendix to complete the following:

    a. Categorize the rocks as mafic or felsic (**Hint:** Use the % dark vs. light minerals in the Appendix).

        i. Rock 2C: _______________________________

        ii. Rock 2D: _______________________________

    b. Name one mineral (other than samples 2A or 2B) that will be abundant in each rock.

        i. Rock 2C: _______________________________

        ii. Rock 2D: _______________________________

    c. Identify each rock by name (assuming these rocks are coarse textured).

        i. Rock 2C: _______________________________

        ii. Rock 2D: _______________________________

PART 3: IGNEOUS ROCKS

The Part 3 rock kit contains 3 rocks. You will use the diagrams at the back of this lab and what you learned in the pre-lab videos to identify these rocks based on composition and texture. *Use your hand lens* to help you see mineral crystals and identify texture.

PROCEDURE

1. Fill out Table 3.3.

   a. **Describe the texture of the sample.** Use these identifiers, which will help you differentiate rock textures according to the table in the Igneous Rock Appendix.

      i.   Glassy = no crystals

      ii. Fine = crystal not visible except with hand lens

      iii. Coarse = crystals visible with the naked eye

      iv. Very coarse (pegmatitic) = crystals larger than a few millimeters

      v.   Vesicular = glass or fine with air pockets

      vi. Porphyritic = "chocolate chip cookie" → fine crystal matrix with phenocrysts

   b. **Estimate % dark vs. light minerals** (from Igneous Rock Appendix).

   c. **Write other observations you've made about the rock that may help with ID.**

   d. **Use the chart in the Appendix to identify the rocks by name.**

DATA COLLECTION

Igneous rock identification.

TABLE 3.3

SAMPLE	TEXTURE	% DARK VS. LIGHT MINERALS OR "NO MINERALS"	OTHER OBSERVATIONS	ROCK NAME
3A				
3B				
3C				

1. You saw 7 igneous rocks total in this lab—2 in Part 1, 2 in Part 2, and 3 in Part 3.

   a. List the felsic igneous rock(s) by sample number. _______________________

   b. List the intermediate igneous rock(s) by sample number. _______________________

   c. List the volcanic (extrusive) rock(s) by sample number (count porphyritic rocks as extrusive). _______________________

   d. List the plutonic (intrusive) rock(s) by sample number. _______________________

PART 4: CLASSIFYING METAMORPHIC ROCKS

The rock kit contains 4 metamorphic rocks. Use the appendix to ID them.

PROCEDURE

1. Fill out Table 3.4.

   a. Describe each rock's texture and color and/or other identifying characteristics.

   b. Use a hand lens to inspect minerals within the rock and suggest at least one mineral that is present.

   c. Identify a potential parent rock.

   d. Identify the metamorphic rock by name.

DATA COLLECTION

Metamorphic rock identification.

TABLE 3.4

SAMPLE	IDENTIFYING CHARACTERISTICS	MINERALS PRESENT	PROTOLITH/ PARENT ROCK	ROCK NAME
4A				
4B				
4C				
4D				

2. You saw 6 metamorphic rocks total in this lab—2 in Part 1 and 4 in Part 4.

   a. List the foliated samples by sample number. _______________________

   b. Put foliated sample numbers in order by metamorphic grade (low → high).

   ____

   c. List the unfoliated sample(s) by sample number. _______________________

APPENDIX 1: IGNEOUS ROCK IDENTIFICATION CHART

Igneous rocks are classified by mineral composition and texture.

1. Compare the color of your igneous rocks to Figure 3.2 to determine the overall mineral composition.

   a. If <~20% dark minerals, **felsic**

   b. If 20–50% dark minerals, **intermediate**

   c. If +50% dark minerals, **mafic** to ultramafic

      i. If dark and greenish, **ultramafic**

2. Find the section on the next page that corresponds to the overall mineral composition of the rock. Use this chart to find other minerals present in the sample.

3. Identify your rocks by finding the name on the upper chart that is in the same column as the overall mineral composition and in the same row as the "coarse" texture.

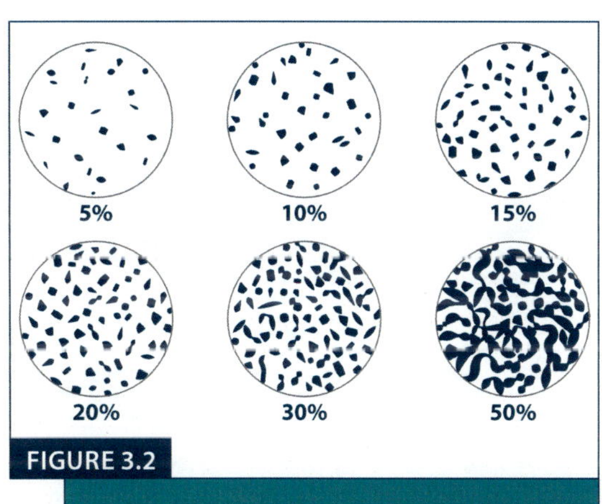

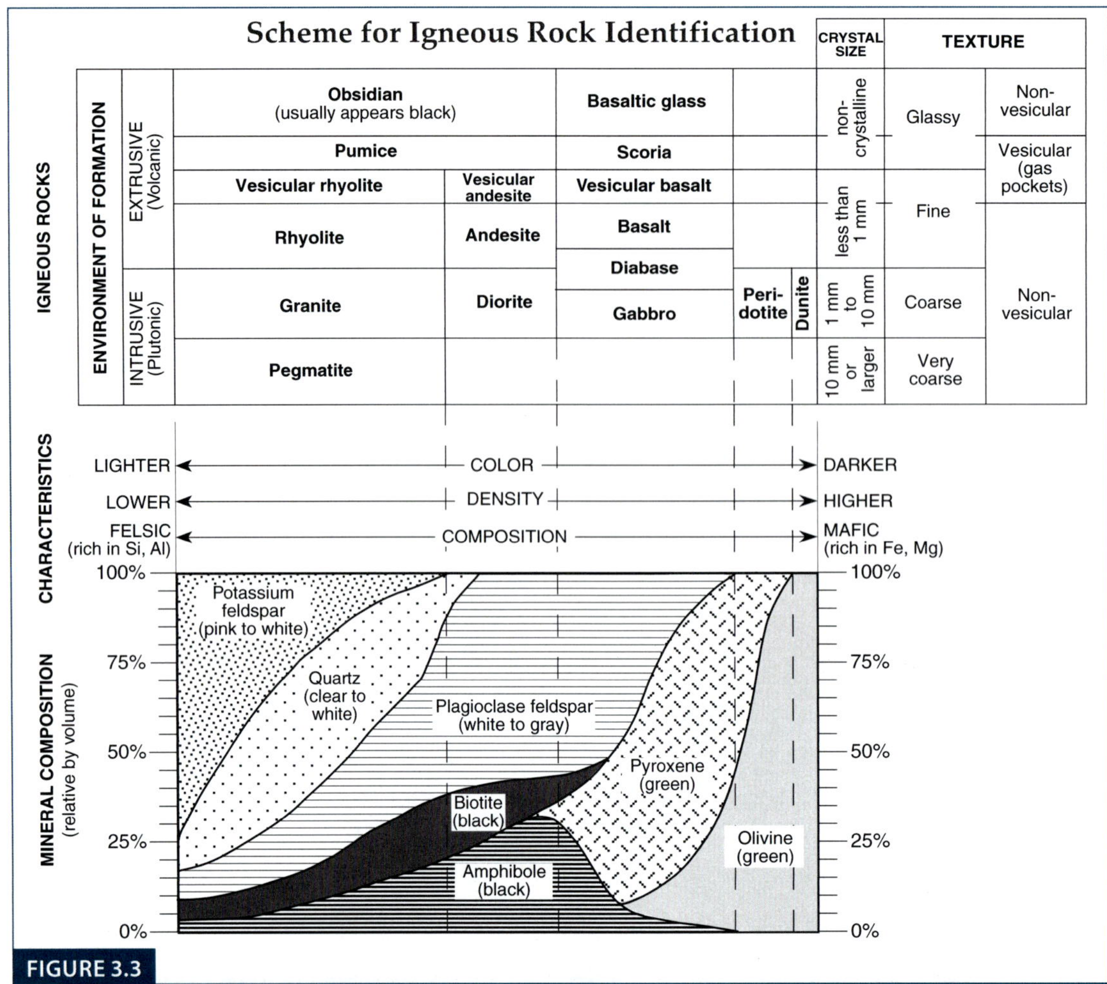

Scheme for igneous rock identification. Source: from the Earth Science Reference Tables from the NY regents exam. http://www.p12.nysed.gov/assessment/reftable/earthscience-rt/esrt2011-engr.pdf.

APPENDIX 2: METAMORPHIC ROCK IDENTIFICATION

Scheme for Metamorphic Rock Identification

TEXTURE		GRAIN SIZE	COMPOSITION	TYPE OF METAMORPHISM	COMMENTS	ROCK NAME	MAP SYMBOL
FOLIATED	MINERAL ALIGNMENT	Fine	MICA / QUARTZ / FELDSPAR / AMPHIBOLE / GARNET / PYROXENE	Regional (Heat and pressure increases)	Low-grade metamorphism of shale	**Slate**	
		Fine to medium			Foliation surfaces shiny from microscopic mica crystals	**Phyllite**	
					Platy mica crystals visible from metamorphism of clay or feldspars	**Schist**	
	BAND-ING	Medium to coarse			High-grade metamorphism; mineral types segregated into bands	**Gneiss**	
NONFOLIATED		Fine	Carbon	Regional	Metamorphism of bituminous coal	**Anthracite coal**	
		Fine	Various minerals	Contact (heat)	Various rocks changed by heat from nearby magma/lava	**Hornfels**	
		Fine to coarse	Quartz	Regional or contact	Metamorphism of quartz sandstone	**Quartzite**	
			Calcite and/or dolomite		Metamorphism of limestone or dolostone	**Marble**	
		Coarse	Various minerals		Pebbles may be distorted or stretched	**Metaconglomerate**	

FIGURE 3.4

Scheme for metamorphic rock identification. Identification table for metamorphic rocks using grain size, texture, and composition. Source: from the Earth Science Reference Tables from the NY regents exam. http://www.p12.nysed.gov/assessment/reftable/earthscience-rt/esrt2011-engr.pdf.

FIGURE 3.5

Scheme for metamorphic rock identification. Identification flow chart that can be used to help you name metamorphic rock samples. Modified from Kevin Stewart (UNC)

4

SEDIMENTARY ROCKS

Name: _______________________________   Section: ___________   Date: _________

Some sections adapted from NC State University's MEA 110 Rocks lab. Appendix 2 from NCSU.

PREFACE

INTRODUCTION

Today, we're giving you the supplies you need to help you answer the following question:

- **How can observations of sediments or sedimentary rocks help you determine how and where they formed?**

LEARNING OBJECTIVES

Students will be able to …

- Compare and contrast clastic, chemical, and biochemical sedimentary rocks.
- Describe the environment of formation for a sedimentary rock based on physical characteristics such as grain size, angularity, and sorting.
- Identify some common sedimentary rocks, including conglomerate, sandstone, shale, limestone, rock salt, and coal.
- Correlate mineral composition of a sedimentary rock with exposure to weathering.

MATERIALS

- Hand lens
- Rocks samples
- Sediment vials
- Mineral ID kit

BEFORE YOU BEGIN

Answer the following questions about sedimentary rocks. Use your pre-lab notes and work together.

1. How does a well-sorted sedimentary rock differ from a poorly-sorted sedimentary rock? What can you infer about the environment of formation?

2. How would the environment of formation for a sedimentary rock with fine, rounded grains differ from that of a sedimentary rock with coarse, angular grains?

3. What does crystalline mean? What type(s) of sedimentary rocks are crystalline? Why?

4. What can you learn about the history of a sedimentary rock by observing the minerals present in the rock? How is minerology related to weathering?

PART 1: SEDIMENTS

You have 3 vials of sediments. Pour some of the sediments from one beaker onto a white sheet of paper on the table, and use the tools provided to fill out the table.

PROCEDURE

1. Fill out Table 1.

    a. Characterize the size(s) of the clasts/sediments using Figure 4.1 below.

    b. Describe the sediments as "well-sorted," "moderately sorted," or "poorly-sorted." Use Figure 4.2 to help you and remember that sorting measures how close to the same size the sediments are.

    c. Describe the sediments as "angular," "sub-rounded," or "rounded," according to Figure 4.3.

    d. Identify minerals present in the sediments. Note that difficult-to-weather minerals, such as quartz, may make up the entirety of some clastic sedimentary rocks, whereas easily-weathered minerals are often found within mineralogically diverse sediment.

DATA COLLECTION

Sediment characterization.

TABLE 4.1

SAMPLE	SEDIMENT SIZE(S)	SORTING	ROUNDING	MINERALS PRESENT
1				
2				
3				

Sediment size classification.

TABLE 4.2

	0.0004 mm	0.063 mm	0.25 mm	0.5 mm	2 mm	4 mm	16 mm
Sediment Name	Clay	Silt	Fine Sand	Medium Sand	Coarse Sand	Granules	Pebbles	Cobbles
Rock Name	Mudstone		Sandstone			Conglomerate or Breccia		
	Claystone	Siltstone						

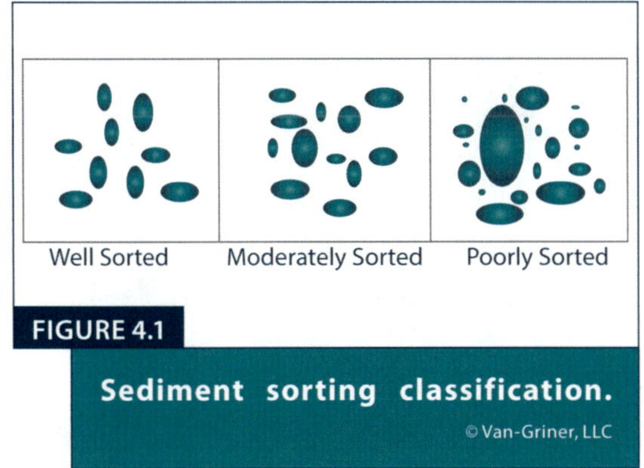

FIGURE 4.1

Sediment sorting classification.
© Van-Griner, LLC

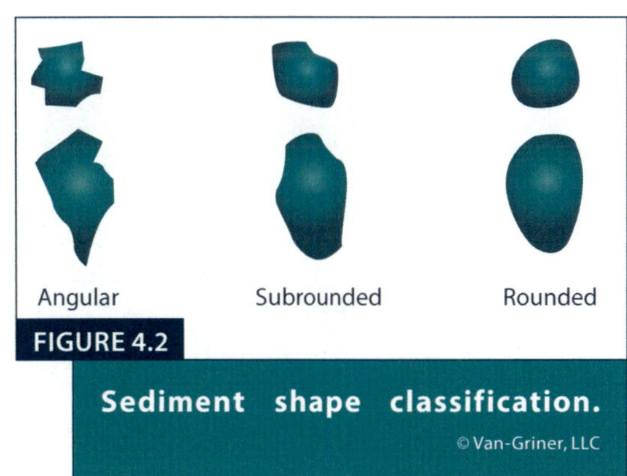

FIGURE 4.2

Sediment shape classification.
© Van-Griner, LLC

PART 2: DEPOSITIONAL ENVIRONMENTS

Differences in sediment shape, size, and sorting tell us about their transport to and the characteristics of their depositional environment (where sediments are deposited).

In general, the larger, more poorly-sorted, and angular sediments are, the closer they were deposited to their source. Smaller, more rounded, and better sorted sediments were typically transported a long way or for a long time before deposition.

PROCEDURE

1. Compare and contrast the geologic history of the sediments in vials 1 and 2, assuming they both started out as granite.

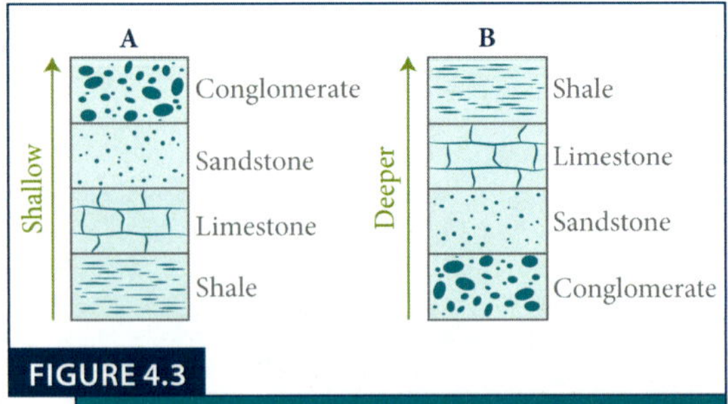

FIGURE 4.3

How sedimentary rock layers reflect depositional environment. If sea level changes over time, then we may see sequences of sedimentary layers that illustrate this change. Figure A and B show how rock layers may appear if sea level became shallower over time (A), or deeper over time (B).

© Van-Griner, LLC

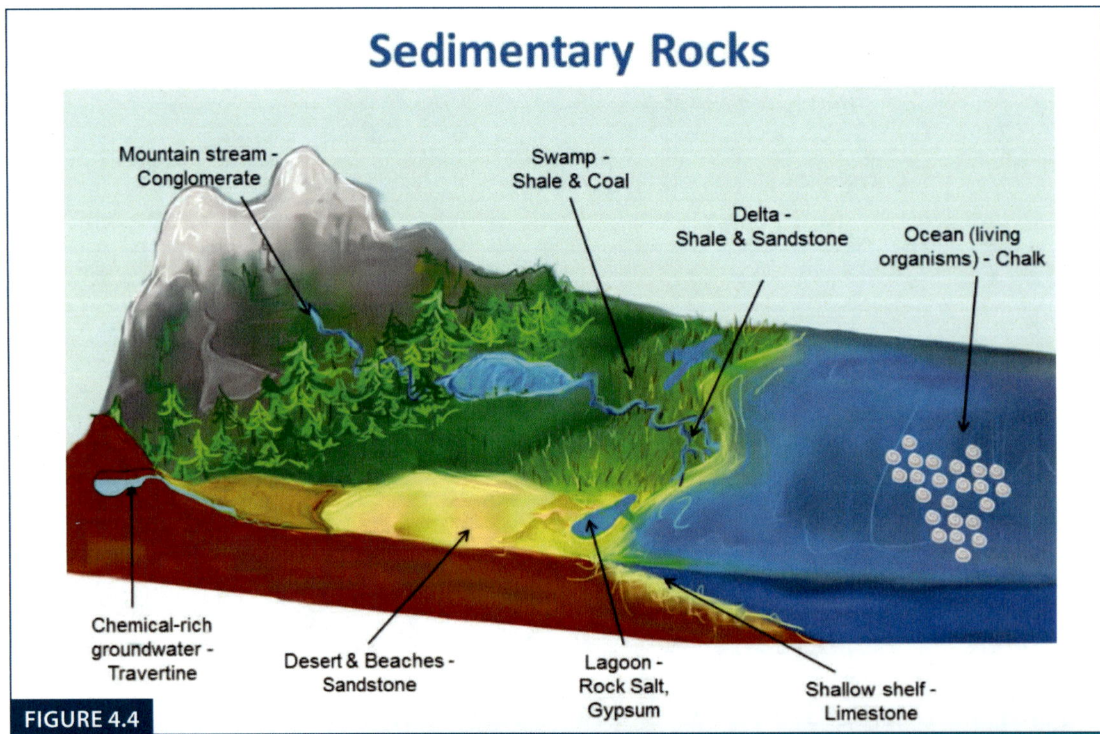

FIGURE 4.4

How sedimentary rock layers reflect depositional environment. This shows typical environments of formation for sedimentary rocks (where conditions are conducive to sediments being deposited and turned into rock). Along a beach, sand is deposited, which can create sandstone; clays are deposited in deeper waters forming shale; calcareous clays are deposited around reefs forming limestone; and small rocks / pebbles are deposited at rivers and mountain bases forming conglomerates.

Source: David McConnell from NC State

2. In the blanks provided (below), sketch the strata (sedimentary rock layers) in an area that started out as a beach, but then underwent a rise in sea level until it was submerged in the deep sea, followed by a decline in sea level that made it so that the elevation matched that of a mountain stream. Use Figure 4.4 as an example, and use the same patterns used in Figure 4.4 to illustrate the rock layers.

3. The bottom sandstone layer may have been cemented together by something that chemically weathered from a layer of sediment that was laid on top of it. **What mineral could be cementing the sandstone together? Justify your answer.** (*Hint:* refer to pre-lab reading for ideas.)

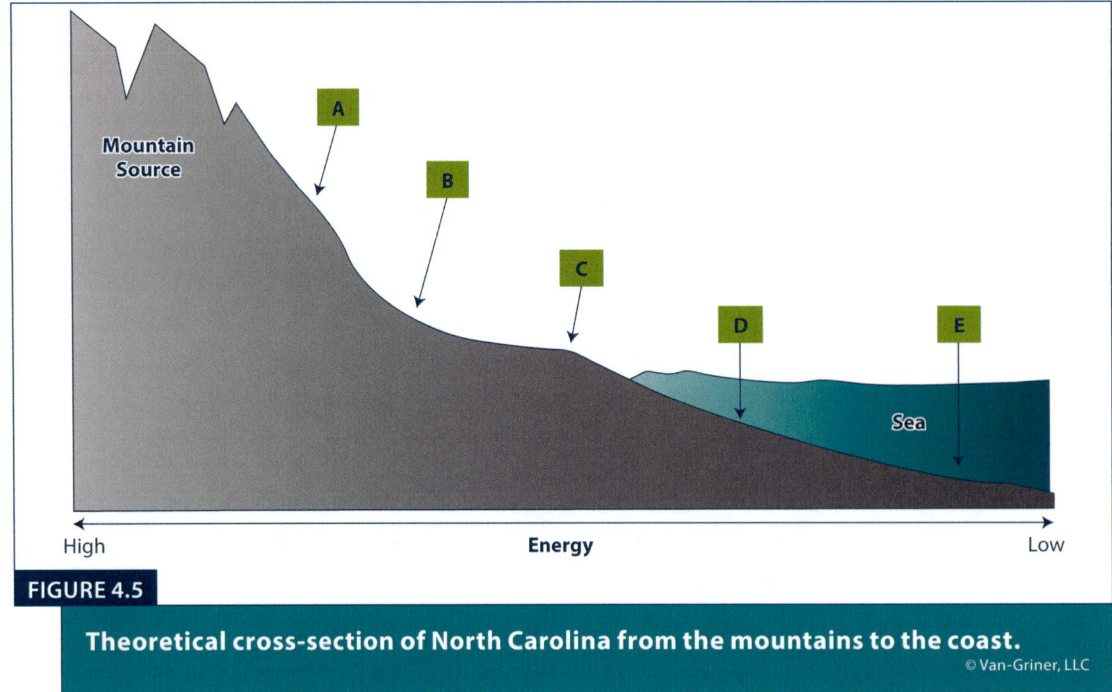

FIGURE 4.5

Theoretical cross-section of North Carolina from the mountains to the coast.

© Van-Griner, LLC

4. Match the rock types below with a letter on Figure 4.5 illustrating the most likely environment of deposition for each rock.

    a. Breccia _________________

    b. Sandstone _________________

    c. Siltstone _________________

    d. Conglomerate _________________

    e. Shale _________________

PART 3: CLASSIFYING SEDIMENTARY ROCKS

The rock kit contains 8 sedimentary rocks. Use the appendix to help you ID them.

PROCEDURE

1. Fill out Table 5 (numbering follows rock tables from previous lab):

    a. Describe each rock's texture and color and/or other identifying characteristics in Column 1.

    b. Categorize the rocks as clastic, chemical, or biochemical/bioclastic.

    c. Use a hand lens to inspect minerals within the rock and suggest at least one mineral that is present.

    d. *Identify* the sedimentary rock by name.

DATA COLLECTION

Identifying sedimentary rocks (numbering of tables follows previous lab)

TABLE 4.3				
SAMPLE	KEY CHARACTERISTICS	ROCK TYPE (CLASTIC, CHEMICAL, BIOCHEMICAL/ BIOCLASTIC)	MINERALS PRESENT, IF ANY	NAME OF ROCK
5A				
5B				
5C				
5D			Clay minerals	
5E				
5F				
5G			No minerals	
5H				

PART 4: ROCK REVIEW

PROCEDURE

1. Identify all the rocks by name, and label rock type for each column. Try to do it without referring back to the answers in your other labs (though you should use the ID charts provided).

TABLE 4.4

SAMPLE ID	ROCK ID	SAMPLE ID	ROCK ID	SAMPLE ID	ROCK ID
Rock type		Rock Type		Rock Type	
1A		1C		5A	
1B		1D		5B	
2C		4A		5C	
2D		4B		5D	
3A		4C		5E	
3B		4D		5F	
3C				5G	
				5H	

PART 5: ON YOUR OWN FOR EXTRA PRACTICE (NOT REQUIRED FOR COMPLETION GRADE)

Look at the labelled rocks in the rock garden and answer the following questions. Photos of the labelled rock garden rocks will be available online, but looking at pictures of rocks is much different than looking at the actual rocks. There are also review questions on these rocks in the "tests and quizzes" section of Sakai. When you submit the quiz, you will get the correct answers. You can do the quiz multiple times, as the questions change each time, to help you study.

PROCEDURE

1. Is rock #4 igneous or metamorphic? How can you tell?

    a. Metamorphic, because it has foliations

    b. Igneous, because there are visible mineral crystals

    c. Igneous, because it formed from magma

    d. Metamorphic, because there are no visible mineral crystals

2. What can you conclude about the mineral composition and cooling history of Rock #2?

    a. This is a felsic, extrusive rock.

    b. This is a mafic, extrusive rock.

    c. This is a felsic, intrusive rock.

    d. This is a mafic, intrusive rock.

3. What is the environment of deposition for this Rock #8?

    a. Beach or desert

    b. Swamp or deep sea

    c. Mountain stream

    d. Shallow marine

4. What type of rock is Rock #7?

    a. Biochemical sedimentary

    b. Chemical sedimentary

    c. Clastic sedimentary

5. Identify Rock #1.

    a. Granite

    b. Rhyolite

    c. Gabbro

    d. Basalt

6. Identify Rock #9.

    a. Shale

    b. Fossiliferous limestone

    c. Sandstone

    d. Conglomerate

APPENDIX 1: SEDIMENTARY ROCK IDENTIFICATION TABLE

INORGANIC LAND-DERIVED SEDIMENTARY ROCKS					
TEXTURE	**GRAIN SIZE**	**COMPOSITION**	**COMMENTS**	**ROCK NAME**	**MAP SYMBOL**
Clastic (fragmental)	Pebbles, cobbles, and/or boulders embedded in sand, silt, and/or clay	Mostly quartz, feldspar, and clay minerals; may contain fragments of other rocks and minerals	Rounded fragments	Conglomerate	
			Angular fragments	Breccia	
	Sand (0.006 to 0.2 cm)		Fine to coarse	Sandstone	
	Silt (0.0004 to 0.006 cm)		Very fine grain	Siltstone	
	Clay (less than 0.0004 cm)		Compact; may split easily	Shale	

CHEMICALLY AND/OR ORGANICALLY FORMED SEDIMENTARY ROCKS					
TEXTURE	**GRAIN SIZE**	**COMPOSITION**	**COMMENTS**	**ROCK NAME**	**MAP SYMBOL**
Crystalline	Fine to coarse crystals	Halite	Crystals from chemical precipitates and evaporites	Rock salt	
		Gypsum		Rock gypsum	
		Dolomite		Dolostone	
Crystalline or bioclastic	Microscopic to very coarse	Calcite	Precipitates of biologic origin or cemented shell fragments	Limestone	
Bioclastic		Carbon	Compacted plant remains	Bituminous coal	

FIGURE 4.6

Scheme for sedimentary rock identification. Source: http://www.p12.nysed.gov/assessment/reftable/ earthscience-rt/esrt2011-engr.pdf.

APPENDIX 2: SEDIMENTARY ROCK IDENTIFICATION FLOW CHART

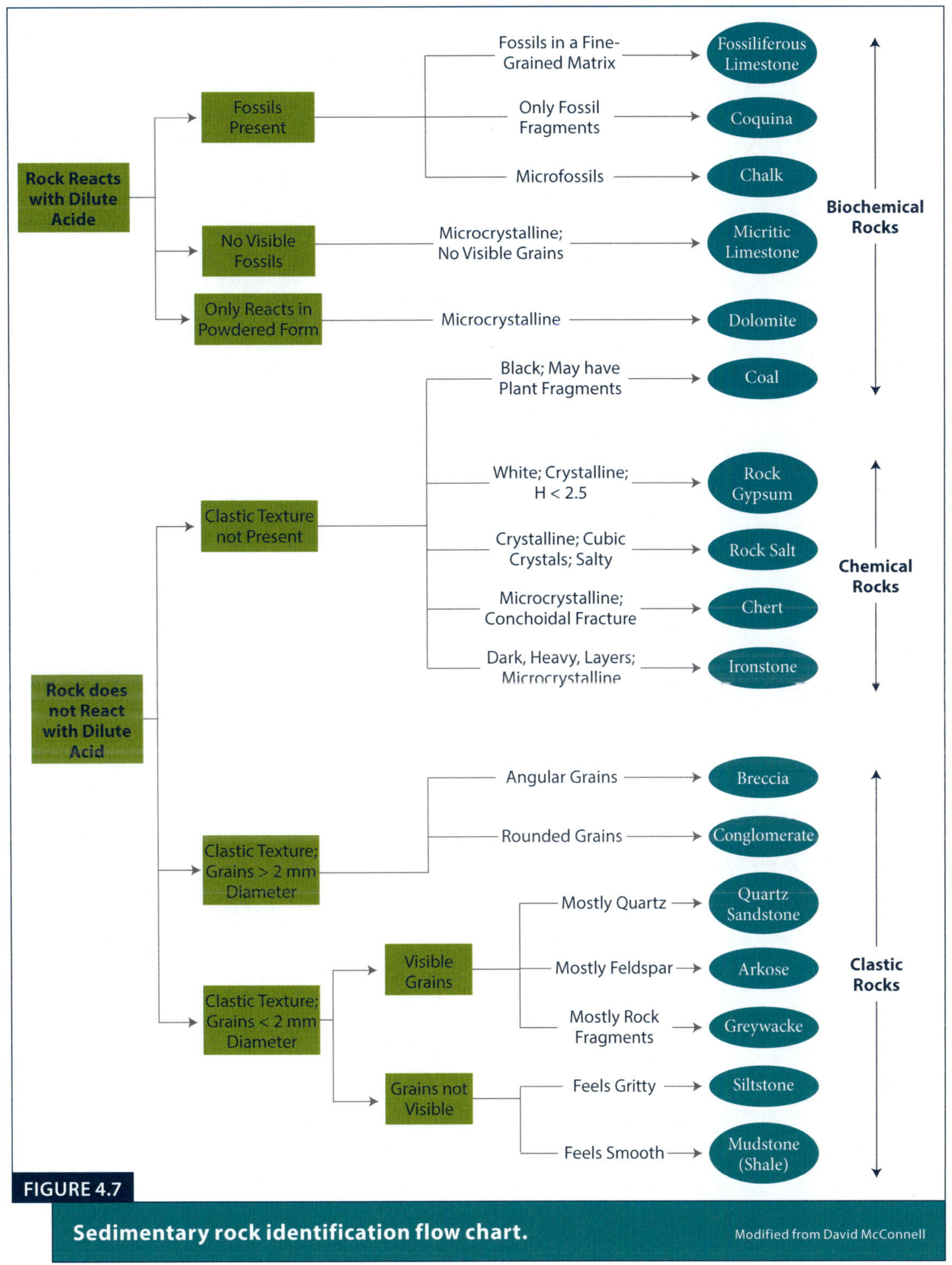

FIGURE 4.7

Sedimentary rock identification flow chart.

Modified from David McConnell

5

TOPO AND GEOLOGIC MAPS

Name: _________________________   Section: __________   Date: ________

PREFACE

INTRODUCTION

Some of the **questions** asked in this lab are: **"What distance and elevation do Dr. Plenge and Dr. X have to cover on their respective bike rides to campus?"** and **"Is land elevation in Chapel Hill related to rock type, and if so, how?"** Feel free to come up with a question of your own as well!

LEARNING OBJECTIVES

Students will be able to …

- ⊙ Collect, analyze, and interpret data pertaining to the study of Earth.
- ⊙ Estimate the elevation of a location on a topographic map that is not on a contour line.
- ⊙ Calculate the gradient (slope) of the land by measuring elevation and distance on a topographic map.
- ⊙ Identify rock units on a geologic map, and determine absolute and relative ages of geologic units on a geologic map using the "Divisions of Geologic Time."
- ⊙ Compare a geologic and topographic map of an area to generate hypotheses about how rock type and elevation are related.

MATERIALS

- ⊙ Topographic & Geologic Maps of Chapel Hill, Divisions of Geologic Time (in class & on Sakai)
- ⊙ Rock samples, rulers, string (in class)
- ⊙ Lab handout

TOPOGRAPHIC AND GEOLOGIC MAPS OF CHAPEL HILL ──

Please read carefully! Use the topographic and geologic maps of Chapel Hill to answer the following questions. *Italicized* "hints" give further instruction for specific questions—read them carefully. Make sure to look at the contour interval and scale bars on the map. Today, use feet for elevation and miles for distance.

Note: Only use wet-erase markers to make marks on the map.

1. Dr. Plenge rides her bike to school every day. Her bike ride starts at the intersection of Simpson St. and W Main St. in Carrboro. A blue dot on the topographic map shows this location.

   a. What is the elevation of her starting point? _________________________________

   Hint: The contour interval is shown at the bottom of the map. Elevation decreases approaching a stream. You can write "between X and Y elevation" if the point does not sit on a contour line.

2. Her bike ride ends at Mitchell Hall, which is at the intersection of 86 & McCauley St./South Rd. A purple dot on the map shows this location.

   a. What is the elevation of her destination? _____________________________

3. The bike route is W. Main St. → E. Main St. → W. Rosemary Ave. → 86 → 86 & McCauley St./South Rd.

   a. How far does she have to bike, in miles? _______________________________

   Hint: The visual scale is shown at the bottom of the map. The horizontal distance is not related to the contour lines. You can use a string—trace it along the route and then straighten and measure with scale—or estimate with a ruler.

4. On the first part of her ride, she has to climb a hill on W. Main St.

   a. The top of the first hill is near the intersection of W. Main St. and __________ Rd.

   b. The elevation of the peak of that hill is greater than __________ ft, but less than __________ ft.

   Hint: The top of the hill must be greater in elevation than the contour lines going around the peak.

5. Move to the geologic map and find Dr. Plenge's bike route there. Note the start and finish points have been marked with dots.

    a. How many geologic units does she travel through? _______________

    b. List the codes for each unit here: _______________

    c. List the rock types that make up the bedrock along her route. _______________

6. The first letter of each geologic unit code refers to the age of the unit. Using Appendix 1 answer these questions.

    a. What is the age (range) of the rock unit she starts on? _______________

    b. What is the age (range) of the rock unit she ends on? _______________

    c. What does "Ma" stand for? _______________

7. Go back to the topographic map. Professor X lives near Weaver Mine Trail, off Raleigh Rd. just east of 15/501. His commute starts at the green dot on your map. Assume he takes Raleigh Rd. west, where it turns into South Rd. and brings him to Mitchell Hall at the intersection of South Rd./McCauley and 86 (the purple dot again).

    a. What's the starting elevation for his commute? _______________

    b. How does this compare to Dr. Plenge's starting elevation? Be specific. _______________

    c. What is the total distance for his commute? _______________

    d. What is the distance between 15/501 and Country Club Rd. along South Rd?

    e. What is the elevation change between 15/501 and Country Club Rd.? Show your work.

    f. What is the slope along *that stretch of the commute?* (between 15/501 and Country Club Rd.) Show your work and express your answer as feet/mile. *(Hint: To calculate slope, divide the elevation change by the horizontal distance. Use feet for elevation and miles for distance to get the correct units).* _______________

    g. Slope is often calculated as percent grade. To calculate a percent grade, your elevation and horizontal distance must be in the same units (e.g., both in feet, or both in miles), and then multiply by 100. **What is the percent grade along this stretch of the commute?** (*Hint: 1 mile = 5280 feet*) _______________

    h. Compare your answer to the grade shown on the "map my ride" profile. If you're not in class to see this or would like to double check your work, you can do so by going to: *https://www.mapmyride.com/routes/my_routes/.* Hit "login" in the upper right, and login using: *mfplenge@unc.edu* for the email and GEOL101L for the password. *If your answer is not within 0.4% of the number on this program, you should redo until you get it closer!*

8. Practice finding the slopes of other areas! Choose 2 other roads and find the slope (ft/mile) of a measured distance along it.

Place 1

    a. Starting place 1 (intersection): _______________ ; elevation: _______________

    b. Ending place 1 (intersection): _______________ ; elevation: _______________

    c. Distance between them: _______________ ; change in elevation between them: _______________

    d. Slope: _______________

Place 2

    e. Starting place 2 (intersection): ________________ ; elevation: ________________

    f. Ending place 2 (intersection): ________________ ; elevation: ________________

    g. Distance between them: ________________ ; change in elevation between them: ________________

    h. Slope: ________________

9. Look at geologic map and find Dr. X's route.

    a. How many geologic units does this route cross? ________________

    *(**Hint:** Include top of hill.)*

    b. List the codes for each unit here: ________________

    c. What rock type is seen here but not on Dr. Plenge's bike route? ________________

    d. What is the age (range) of the rock unit Dr. X starts on? ________________

    e. What is the age (range) of the rock unit Dr. X ends on? ________________

    f. Describe how Dr. Plenge's and Dr. X's bike rides are different in terms of total distance and elevation change (who has a harder bike ride?) and in terms of geology (how do the rocks and ages compare)?

    g. Find at least 2 other elevations on the map that are at the same elevation as the highest point of Dr. X's commute. How do the rock types and ages compare? Fill out the chart below:

LOCATION (AS LAT/LONG OR NEAREST INTERSECTION)	ROCK TYPE	ROCK AGE

    h. Did the table illustrate any patterns in topography and rock age? Justify your answer.

10. Each "U" on the geologic map indicates a rock unit that has been uplifted compared to surrounding units, and "D" indicates relative movement down. The dotted lines represent the breaks (faults) in the rock along which there has been movement. Locate the "U" and the two "D" areas at Weaver Mine Trail.

    a. Explain how what you see in the geologic map relates to the topography of the area around Weaver Mine Trail.

    b. What type of rock is the uplifted geologic unit? _______________________

    c. What type of rock is the downthrown geologic unit? _______________________

    d. Which rock is older, the one at the top of the hill or the one at the bottom? _______________________

11. Develop a hypothesis explaining your answer to 10d.

12. Examine the rock samples given that correspond to the different types of rocks in Chapel Hill, Carrboro, and Durham. Find the samples that correspond to the rocks in #10b and #10c.

    a. Which rock do you think would be the easiest to weather/erode? Why?

    b. The Trcs/sil unit used to be at the top of the Weaver Mine Trail hill but is no longer there. Why not (**Hint:** *Think back to your pre-lab reading from Lab 4*)?

    c. Do you think younger rock units may have once sat on top of the rock units Dr. Plenge bikes over on her route? Explain your reasoning.

d. Does thinking about rock types and weathering/erosion change your hypothesis from 10? If your hypothesis did not include weathering/erosion, write a new hypothesis **here** that explicitly includes weathering/erosion. If your hypothesis **did** include weathering and erosion, then you should write a new scientific question about Chapel Hill you've developed by looking at the maps!

APPENDIX 1

Divisions of Geologic Time—
Major Chronostratigraphic and Geochronologic Units

Introduction.—Effective communication in the geosciences requires consistent uses of stratigraphic nomenclature, especially divisions of geologic time. A geologic time scale is composed of standard stratigraphic divisions based on rock sequences and calibrated in years (Harland and others, 1982). Over the years, the development of new dating methods and refinement of previous ones have stimulated revisions to geologic time scales.

Since the mid-1990s, geologists from the U.S. Geological Survey (USGS), State geological surveys, academia, and other organizations have sought a consistent time scale to be used in communicating ages of geologic units in the United States. Many international debates have occurred over names and boundaries of units, and various time scales have been used by the geoscience community.

New time scale.—Since the publication by the USGS of the 7th edition of "Suggestions to Authors" (STA7; Hansen, 1991), no other time scale has been officially endorsed by the USGS. For consistency purposes, the USGS Geologic Names Committee (GNC; see box for members) and the Association of American State Geologists (AASG) developed **Divisions of Geologic Time** (fig. 1). The **Divisions of Geologic Time** is based on the time scale in STA7 (Hansen, 1991, p. 59) and updates it with the unit names and boundary age estimates ratified by the International Commission on Stratigraphy (ICS). Scientists should note that other published time scales may be used, provided that these are specified and referenced (for example, Palmer, 1983; Harland and others, 1990; Haq and Eysinga, 1998; Gradstein and others, 2004). Advances in stratigraphy and geochronology require that any time scale be periodically updated. Therefore, the **Divisions of Geologic Time** is dynamic and will be modified as needed to include accepted changes of unit names and boundary age estimates.

The **Divisions of Geologic Time** shows the major chronostratigraphic (position) and geochronologic (time) units; that is, eonothem/eon to series/epoch divisions. Workers should refer to the ICS time scale (Ogg, 2004) for stage/age terms. Most systems of the Paleozoic and Mesozoic are subdivided into series utilizing the terms "Lower," "Middle," and "Upper." The geochronologic counterpart terms for subdivisions of periods are "Early," "Middle," and "Late." The international geoscience community is applying names to these subdivisions based on stratigraphic sections at specific localities worldwide. All series/epochs of the Silurian and Permian have been named. Although the usage of these names is preferred, "lower/early," "middle," and "upper/late" are still acceptable as informal units (lowercase) for these two systems/periods. Also the Upper Cambrian has been named "Furongian" in the ICS time scale. However, the GNC will not recognize this name and include it in the **Divisions of Geologic Time** until all series/epochs of the Cambrian are named.

Cenozoic.—There has been much controversy related to subdivisions of the Cenozoic, particularly regarding retention or rank of the terms "Tertiary" and "Quaternary." Although some stratigraphers have suggested that these terms be abandoned, the issue remains unresolved. If the terms are retained, there will need to be agreement on the status of the Quaternary as a system/period or subsystem/subperiod. Another controversial issue is the position of the base of the Quaternary: is it at the base of the Pleistocene or within the upper Pliocene? These positions have age estimates of 1.8 Ma and 2.6 Ma, respectively (see box for age terms). Until a decision is made on the subdivisions of the Cenozoic, the **Divisions of Geologic Time** will follow the general structure of the time scale in STA7 (Hansen, 1991) in accepting the use of the terms "Tertiary" and "Quaternary" and the equivalence of the bases of the Quaternary and Pleistocene. The map symbols "T" (Tertiary) and "Q" (Quaternary) have been used on geologic maps for more than a century and are widely used today.

Precambrian.—For many years, the term "Precambrian" was used for the division of time older than the Phanerozoic. For consistency with the time scale in STA7 (Hansen, 1991), the term "Precambrian" is considered to be informal and without specific stratigraphic rank (although it is capitalized).

Map colors.—Geologic maps utilize color schemes based on standards that are related to the time scale. Two different schemes are used, one by the Commission for the Geologic Map of the World (CGMW) and another by the USGS. Colors typically shown on USGS geologic maps have been used in a standard fashion since the late 1800s and recently have been published in the digital cartographic standard for geologic map symbolization (Federal Geographic Data Committee, Geologic Data Subcommittee, 2006). The GNC decided in 2006 that the USGS colors should be used for large-scale and regional geologic maps of the United States. For international maps or small-scale maps (for instance, 1:5 million) of the United States or North America, the GNC recommends the use of the international colors. Specifications for the USGS colors are in Federal Geographic Data Committee, Geologic Data Subcommittee (2006), and those for the CGMW colors are in Gradstein and others (2004).

Acknowledgments.—This Fact Sheet benefited from thoughtful reviews by John Repetski (USGS) and Donald E. Owen (Lamar University, Beaumont, Tex., and North American Commission on Stratigraphic Nomenclature).

By U.S. Geological Survey Geologic Names Committee

Members of the Geologic Names Committee of the U.S. Geological Survey, 2006

Randall C. Orndorff (chair), Nancy Stamm (recording secretary), Steven Craigg, Terry D'Erchia, Lucy Edwards, David Fullerton, Bonnie Murchey, Leslie Ruppert, David Soller (all of the USGS), and Berry (Nick) Tew, Jr. (State Geologist of Alabama).

U.S. Department of the Interior
U.S. Geological Survey

Fact Sheet 2007–3015
March 2007

FIGURE 5.1

Fact Sheet.

Source: USGS Public Domain

Age Terms

The age of a stratigraphic unit or the time of a geologic event may be expressed in years before present (before A.D. 1950). The "North American Stratigraphic Code" (North American Commission on Stratigraphic Nomenclature, 2005) recommends abbreviations for ages in SI (International System of Units) prefixes coupled with "a" for annum: ka for kilo-annum, 10^3 years; Ma for mega-annum, 10^6 years; and Ga for giga-annum, 10^9 years. Duration of time should be expressed in millions of years (m.y.). For example, deposition began at 85 Ma and continued for 2 m.y.

References Cited

Federal Geographic Data Committee, Geologic Data Subcommittee, 2006, FGDC digital cartographic standard for geologic map symbolization: Federal Geographic Data Committee Document Number FGDC–STD–013–2006, 290 p., 2 pls., available online at http://ngmdb.usgs.gov/fgdc_gds/.

Gradstein, Felix, Ogg, James, and Smith Alan, eds., 2004, A geologic time scale 2004: Cambridge, U.K., Cambridge University Press, 589 p., 1 pl.

Hansen, W.R., ed., 1991, Suggestions to authors of the reports of the United States Geological Survey, seventh edition [STA7]: Reston, Va., U.S. Geological Survey, 289 p. (Also available online at http://www.nwrc.usgs.gov/lib/lib_sta.htm.)

Haq, B.U., and Eysinga, F.W.B., van, eds., 1998, Geological time table (5th ed.): Amsterdam, Elsevier, 1 sheet.

Harland, W.B., Armstrong, R.L. Cox, A.V., Craig, L.E., Smith, A.G., and Smith, D.G., 1990, A geologic time scale, 1989: Cambridge, U.K., Cambridge University Press, 263 p.

Harland, W.B., Cox, A.V., Llewellyn, P.G., Picton, C.A.G., Smith A.G., and Walters, R.W., 1982, A geologic time scale: Cambridge, U.K., Cambridge University Press, 131 p.

North American Commission on Stratigraphic Nomenclature, 2005, North American stratigraphic code: American Association of Petroleum Geologists Bulletin, v. 89, no. 11, p. 1547–1591. (Also available online at http://ngmdb.usgs.gov/Info/NACSN/Code2/code2.html.)

Ogg, James, comp., 2004, Overview of global boundary stratotype sections and points (GSSPs): International Commission on Stratigraphy, available online at http://www.stratigraphy.org/gssp.htm.

Palmer, A.R., comp., 1983, The Decade of North American Geology [DNAG] 1983 geologic time scale: Geology, v. 11, no. 9, p. 503–504.

For more information, please contact:

Randall C. Orndorff, U.S. Geological Survey
908 National Center, 12201 Sunrise Valley Drive
Reston, VA 20192
E-mail: rorndorf@usgs.gov

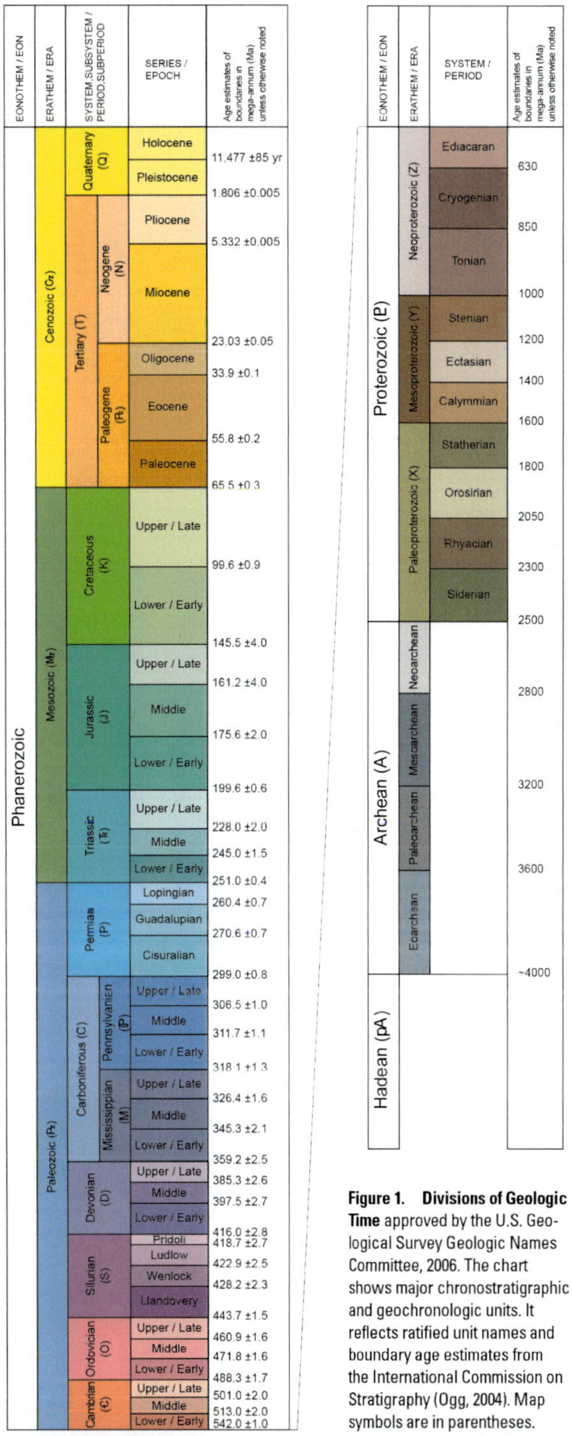

EONOTHEM / EON	ERATHEM / ERA	SYSTEM, SUBSYSTEM / PERIOD, SUBPERIOD		SERIES / EPOCH	Age estimates of boundaries in mega-annum (Ma) unless otherwise noted
Phanerozoic	Cenozoic (Cz)	Quaternary (Q)		Holocene	11,477 ±85 yr
				Pleistocene	1.806 ±0.005
		Tertiary (T)	Neogene (N)	Pliocene	5.332 ±0.005
				Miocene	23.03 ±0.05
			Paleogene (Pg)	Oligocene	33.9 ±0.1
				Eocene	55.8 ±0.2
				Paleocene	65.5 ±0.3
	Mesozoic (Mz)	Cretaceous (K)		Upper / Late	99.6 ±0.9
				Lower / Early	145.5 ±4.0
		Jurassic (J)		Upper / Late	161.2 ±4.0
				Middle	175.6 ±2.0
				Lower / Early	199.6 ±0.6
		Triassic (Tr)		Upper / Late	228.0 ±2.0
				Middle	245.0 ±1.5
				Lower / Early	251.0 ±0.4
	Paleozoic (Pz)	Permian (P)		Lopingian	260.4 ±0.7
				Guadalupian	270.6 ±0.7
				Cisuralian	299.0 ±0.8
		Carboniferous (C)	Pennsylvanian (P)	Upper / Late	306.5 ±1.0
				Middle	311.7 ±1.1
				Lower / Early	318.1 ±1.3
			Mississippian (M)	Upper / Late	326.4 ±1.6
				Middle	345.3 ±2.1
				Lower / Early	359.2 ±2.5
		Devonian (D)		Upper / Late	385.3 ±2.6
				Middle	397.5 ±2.7
				Lower / Early	416.0 ±2.8
		Silurian (S)		Pridoli	418.7 ±2.7
				Ludlow	422.9 ±2.5
				Wenlock	428.2 ±2.3
				Llandovery	443.7 ±1.5
		Ordovician (O)		Upper / Late	460.9 ±1.6
				Middle	471.8 ±1.6
				Lower / Early	488.3 ±1.7
		Cambrian (C)		Upper / Late	501.0 ±2.0
				Middle	513.0 ±2.0
				Lower / Early	542.0 ±1.0

EONOTHEM / EON	ERATHEM / ERA	SYSTEM / PERIOD	Age estimates of boundaries in mega-annum (Ma) unless otherwise noted
Proterozoic (P)	Neoproterozoic (Z)	Ediacaran	
		Cryogenian	630
			850
		Tonian	
	Mesoproterozoic (Y)	Stenian	1000
			1200
		Ectasian	
			1400
		Calymmian	
	Paleoproterozoic (X)	Statherian	1600
			1800
		Orosirian	
			2050
		Rhyacian	
			2300
		Siderian	
			2500
Archean (A)	Neoarchean		
			2800
	Mesoarchean		
			3200
	Paleoarchean		
			3600
	Eoarchean		
			~4000
Hadean (pA)			

Figure 1.   Divisions of Geologic Time approved by the U.S. Geological Survey Geologic Names Committee, 2006. The chart shows major chronostratigraphic and geochronologic units. It reflects ratified unit names and boundary age estimates from the International Commission on Stratigraphy (Ogg, 2004). Map symbols are in parentheses.

FIGURE 5.3

Fact Sheet.

6

BATTLE PARK

Name: _______________________________   Section: ___________   Date: _________

PREFACE

INTRODUCTION

The **question** asked in this lab is **"What causes changes in elevation within a given rock unit (e.g., the Chapel Hill Granite)?"** Feel free to come up with a question of your own as well!

LEARNING OBJECTIVES

COURSE LEARNING OBJECTIVE ADDRESSED

Students will be able to …

- ⊙ Collect, analyze, and interpret data pertaining to the study of Earth.
- ⊙ Generate hypotheses regarding the geologic history of everyday landscapes.

SPECIFIC LEARNING OBJECTIVES FOR "WELCOME TO CHAPEL HILL" LAB

Students will be able to …

- ⊙ Generate a vertical profile by reading contours on a topographic map.
- ⊙ Calculate the vertical exaggeration of a vertical profile/cross-section.
- ⊙ Differentiate chemical and physical weathering and identify examples of chemical and physical weathering in the field.
- ⊙ Define erosion and explain how rock type, vegetation, and other variables control erosion and water flow direction in this area.

SAFETY CONCERNS

- We will be walking down a relatively steep gradient on an unpaved path. Please notify your TA if you have concerns about physically being able to complete the lab.

- Poison ivy, insects, and/or snakes may be present. Please use caution, and immediately point out snakes to your TA for identification and avoidance.

- Wear appropriate clothing, e.g. long pants, closed-toe shoes with good tread, hats, and/or sunscreen.

MATERIALS

- Lab handout, including maps

BEFORE YOU BEGIN

Please read carefully! Meet at Battle Park! Map showing where to meet is on the Sakai page. The class will be waiting at the picnic tables across from the undergraduate admissions building in Battle Park.

PART 1: BEFORE THE WALK

Use the map in Figure 6.3 for the first 6 questions.

1. Using the Battle Park trail map as a guide, draw Battle Branch Creek on the plain contour map in Figure 6.3. Estimate the position as well as you can, and use the "rule of V's" to guide you

2. Indicate the direction of flow of the creek using arrows.

3. Imagine rainfall is occurring only in the 6 areas marked with an "x" on the map. Draw arrows showing the direction surface water flows when falling in those areas. *Hint:* Flow should be perpendicular to contours.

Use the laminated map (see Figure 6.4) for the remaining Part 1 questions.

4. The Rainy Day Trail starts at an elevation of 462 ft. The elevation of the creek at Stop 1 is 372 ft, and the elevation at Stop 2 is ~354 ft.

   a. Faults and other evidence of tectonic displacement are absent within the map area, and bedrock in this area is all the same: granite and granidiorites. Tectonic displacement and bedrock composition often control differences in elevation, but can't be used to explain the elevation differences observed here. In 2–3 words, identify something that **can** account for the elevation changes observed in the park.

b. The distance along the creek between the 2 stops is about 0.06 miles. **What is the stream gradient?** *(**Hint:** Stream gradient is the same as slope (which you calculated last week), or the change in elevation over change in distance.)*

5. We will be walking down the Rainy Day Trail today. After 0.3 miles, the trail runs into the OWASA access road, and we will walk about 0.1 miles further along this road to get to creek stops 1 and 2. A laminated map of Battle Park will be provided (and is available on Sakai), and dashes marking 0.05 mile increments down the trail have been marked.

a. **Create 2 vertical profiles of the hike you'll be taking today.**

   Hint: The bold contour closest to the start of the Rainy Day Trail is 460. Use the contour interval to help you determine the elevation at each marked 0.05 distance down the trail. If you are not on a contour, estimate the elevation as being between the contours on either side of the marked location.

   i. **Graph 1:** Plot the elevation changes that occur while on the Rainy Day Trail (the first 0.3 miles of our hike). Connect the dots with a smooth line to make a profile.

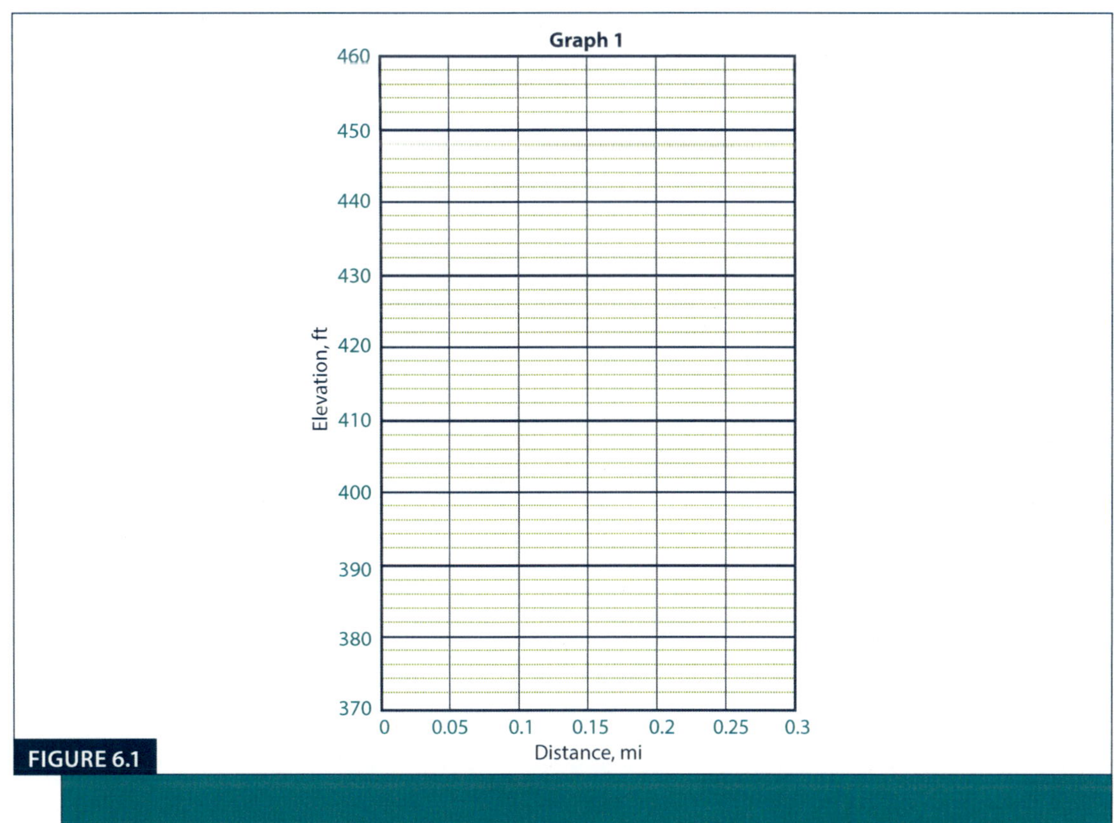

FIGURE 6.1

ii. **Graph 2:** Plot the exact same data on this graph and connect with a smooth line to make a profile. Note that the vertical scale is very different in terms of the number of feet elevation represented by each line on the graph.

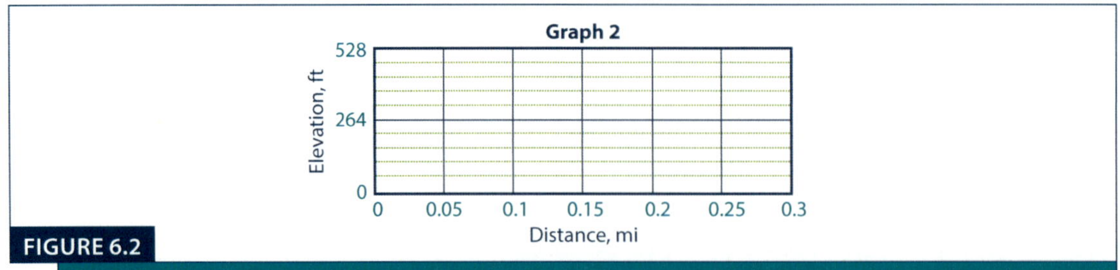

FIGURE 6.2

Example calculation: Vertical Exaggeration of Graph 2. VE = (x-axis scale)/(y-axis scale) = (0.05 miles)/(264 ft) → convert to same units → VE = 264 ft/264 ft = 1 VE = 1. This means the vertical scale is 1x the horizontal scale, so there is NO vertical exaggeration.

b. These graphs show the exact same changes in elevation over distance but using different vertical scales. Graph 2 uses the same scale on the horizontal and vertical axes (note that 264 feet is equal to 0.05 miles). Graph 1 uses vertical exaggeration, e.g. it stretches the y-axis to make the elevation changes easier to visualize. This is done using the equation VE = (x-axis scale)/(y-axis scale), where scale is the distance shown by 1 solid box on the axis. ***Hint:*** *Units must be the same; 0.05 miles = 264 ft. Example calculation can be found below Graph 2.*

i. Calculate the vertical exaggeration (VE) of Graph 1. Show your work.

ii. Try to draw a person, to scale, on each map. Based on the size of that person relative to the slope of the land, decide which graph is a more realistic depiction of the slope of the land?

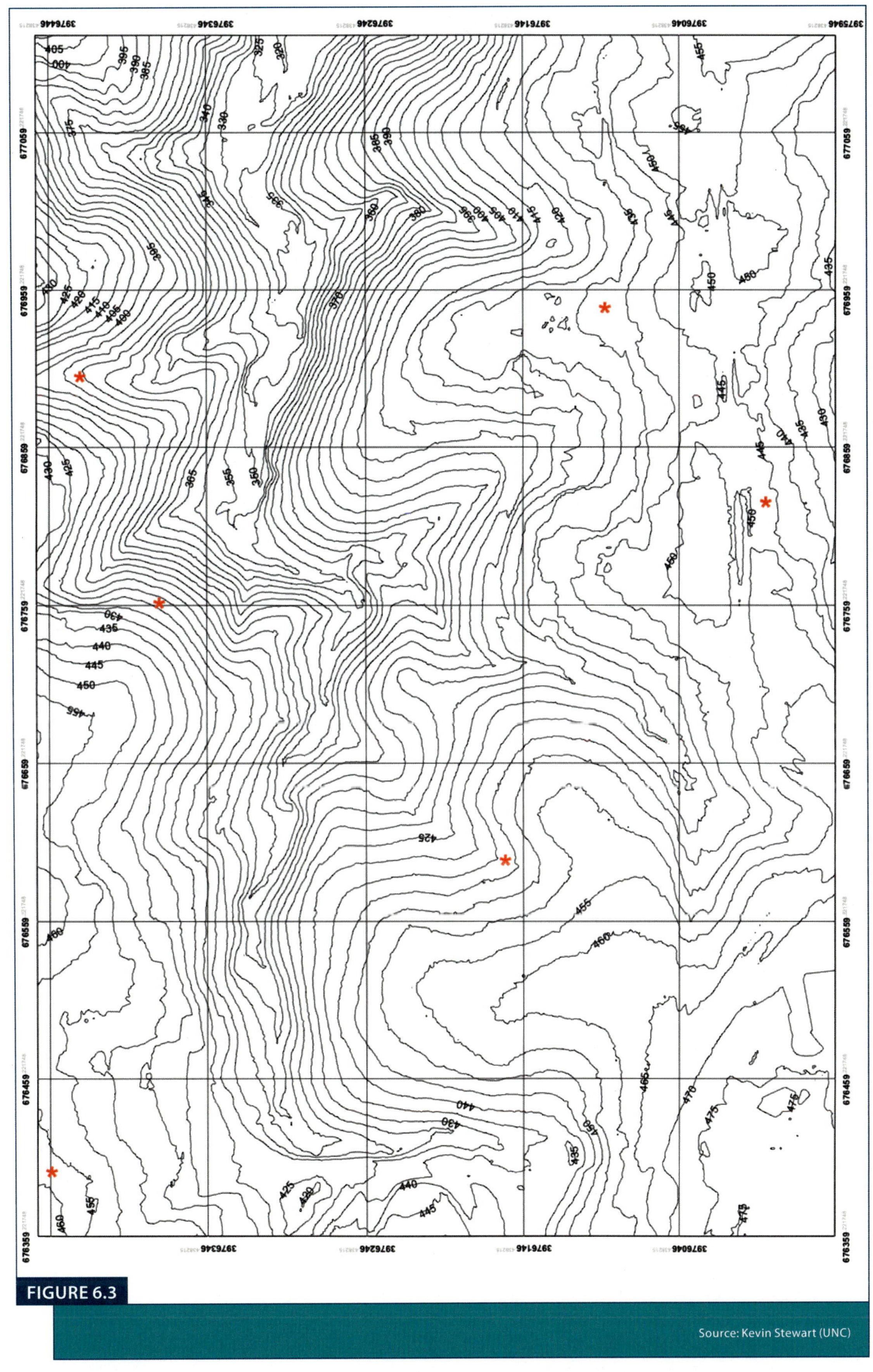

FIGURE 6.3

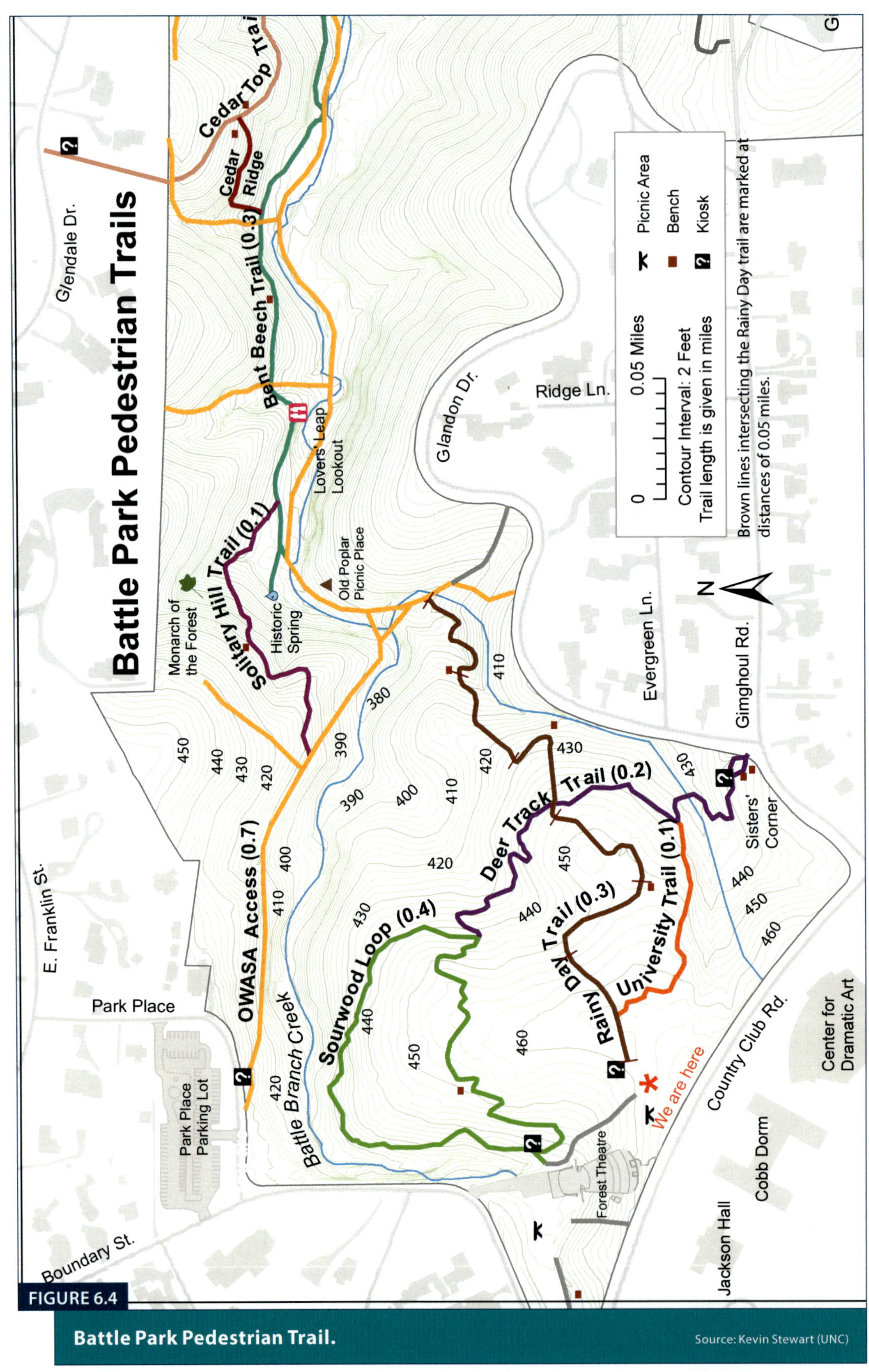

FIGURE 6.4

Battle Park Pedestrian Trail.

Source: Kevin Stewart (UNC)

PART 2: DURING THE WALK

1. Observe the rocks along the path. What kind do you think they are? (**Circle one.**)

    ◎ Igneous   ◎ Sedimentary   ◎ Metamorphic

2. Look at a rock that has been broken open. Does seeing the inside change your answer to #1? How?

3. What can account for the difference in appearance, if any, in the interior and exterior of the rock?

4. The rocks found here are part of the bedrock in this region. Define bedrock and explain how these rocks may have been detached from the bedrock to become loose stones.

5. How do tree roots impact the weathering of bedrock in this area?

6. How do tree roots impact the erosion of bedrock in this area?

7. Creek Stop 1:
    a. Observe the sediments in the bottom of the creek, and describe the size, shape, and composition of the sediments. For composition, note sample color, if sediments look like a single mineral or smaller rock fragments, etc.
        i.   Size: ____
        ii. Shape: ____
        iii. Composition: ____
    b. How do you think anthropogenic (human) influences have contributed to the observed shape/size/composition of the sediments you see here?

8. Walk along the stream until you reach a bend in the stream.

   a. Sketch out how sediment sizes look on the inside vs the outside of the bend in the box to the right.

   b. Based on your pre-activity, how do you expect sediment size to vary with water velocity?

   c. How does water velocity change as you go across the width of the stream?

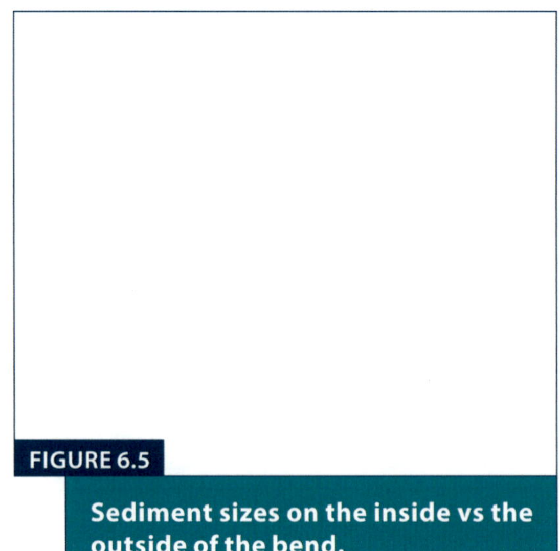

FIGURE 6.5

Sediment sizes on the inside vs the outside of the bend.

9. Creek Stop 2:

   a. Do you think the creek has a greater discharge at Creek Stop 1 or Creek Stop 2? Justify your answer.

   b. Physical and chemical weathering of granite can be seen clearly here. Your TA will break some rock off the wall near the creek at Stop 2. Rub the sediment in your hand, and compare what you feel to what you felt at Creek Stop 1.

      i. How does the rock texture differ from the sediment you felt at the bottom of the stream? **Propose an explanation for any differences you observe.**

      ii. Do you think the source rock for the sediment at Creek Stop 1 is the same type of rock you see here? Justify your answer.

    iii. Battle Branch Creek eventually flows into Bolin Creek, and then to Jordan Lake. How do you think the sediment size, shape, or composition would compare between Jordan Lake and Battle Branch Creek (e.g., what would the sediment in the lake look/feel like)? *Hint: You may want to think about sediments that are missing from this area.*

10. The Rainy Day Trail is within the Chapel Hill granite (Zgr). Battle Branch Creek, shown with the thick blue line on the map, runs through the Chapel Hill Granite towards a Triassic sandstone/siltstone (Trsc/sit) outcrop. **Explain why the creek changes direction when it gets to the tan Trsc/sit outcrop.** *Hint: The stream is still flowing downhill, but the elevation is consistently lower in the sandstone/siltstone as compared to the granite on either side. Why?*

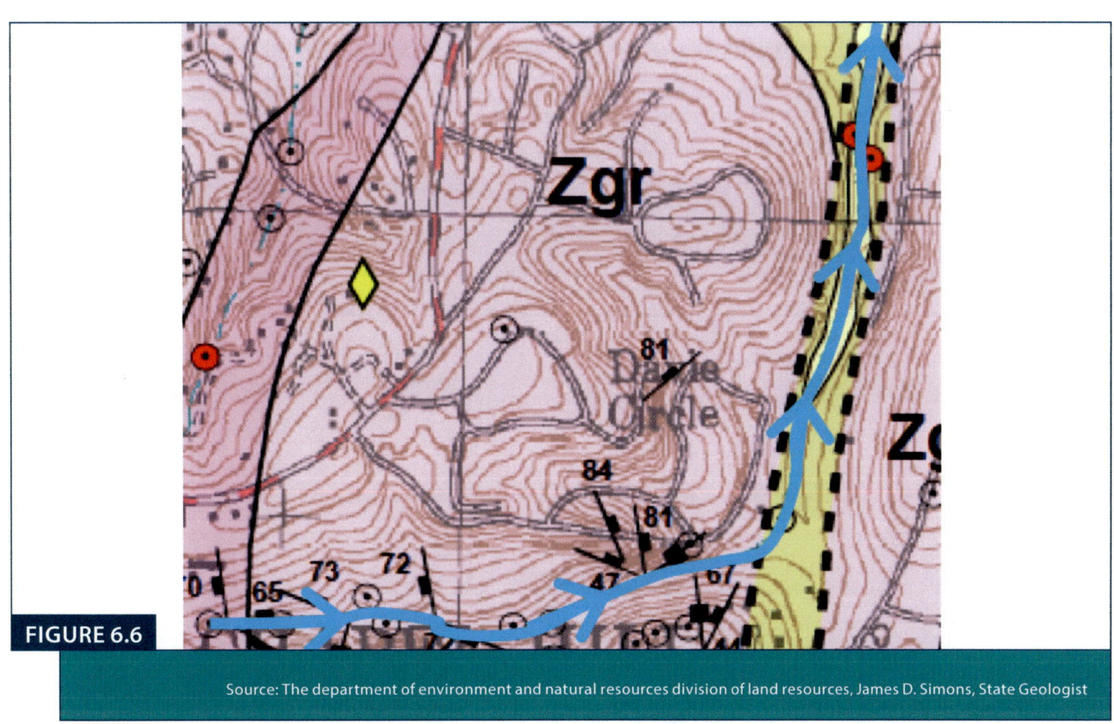

FIGURE 6.6

Source: The department of environment and natural resources division of land resources, James D. Simons, State Geologist

BATTLE PARK— WATER DISCHARGE

Name: _______________________   Section: __________   Date: _________

PREFACE

INTRODUCTION

The **questions** asked in this lab are: **"How does water discharge vary as a river moves downstream?"** and **"How does current water discharge compare to historical water discharge in this stream?"**

LEARNING OBJECTIVES

Students will be able to …

- Collect, analyze, and interpret data pertaining to the study of Earth.
- Describe the method we used to determine stream discharge, and justify methodological decisions including taking depth measurements at different locations, using multiple trials for velocity, and determining the distance between the start and stop line.
- Calculate the discharge of a stream given variables such as depth measurements and velocity measurements.
- Evaluate the accuracy of their own field data by comparing their results to historical results and data collected by their peers.
- Describe how stream discharge changes as you move downstream, and how large precipitation events can alter discharge over time.

SAFETY CONCERNS

- We will be walking down a relatively steep gradient on an unpaved path. Please notify your TA if you have concerns about physically being able to complete the lab.

- Poison ivy, insects and/or snakes may be present. Please use caution.

- Wear appropriate clothing, e.g. long pants, closed-toe shoes with good tread, hats and/or sunscreen.

MATERIALS

- You bring: Lab handout

- TA brings: String, tent stakes, rulers, measuring tape, and ping-pong ball

BEFORE YOU BEGIN

Please read carefully! Meet at Battle Park! Map showing where to meet is on the Sakai page. The class will be waiting at the picnic tables across from the undergraduate admissions building in Battle Park.

PART 1: PRACTICE USING DISCHARGE EQUATION AND CONVERTING INCHES TO FEET

You'll do this portion of the lab sitting at the picnic tables while waiting for everyone to arrive.

Discharge = Area × Velocity × Coefficient

For the coefficient, use 0.9 if the stream bottom along that stretch is mostly sandy; use 0.8 if it is mostly rocky.

1. Location 1 and location 2 are 2 feet apart. At both locations, the stream has a uniform depth of 8 inches; and a width of 6 feet. It takes a ping pong ball 45 seconds to float from location 1 to location 2.

   a. What is the cross-sectional area of the stream in feet? Show your work. (***Hint: To convert inches to feet, divide by 12.***)

   b. What is the velocity of the stream? What is the velocity of the stream? Velocity can be calculated using the equation: Velocity = Distance/Time. Show your work.

c. What is the discharge of the stream, assuming it has a rocky bottom? Show your work.

d. What is the discharge of the stream, assuming it has a sandy bottom? Show your work.

e. Do large or small sediments slow the stream down more? (**Circle one.**)

2. You go to location X, which is upstream from locations 1 and 2. Here, the stream is still 6 feet wide, but the depth varies (see diagram). To determine the depth, you measure the depth of the stream in three different locations and average them.

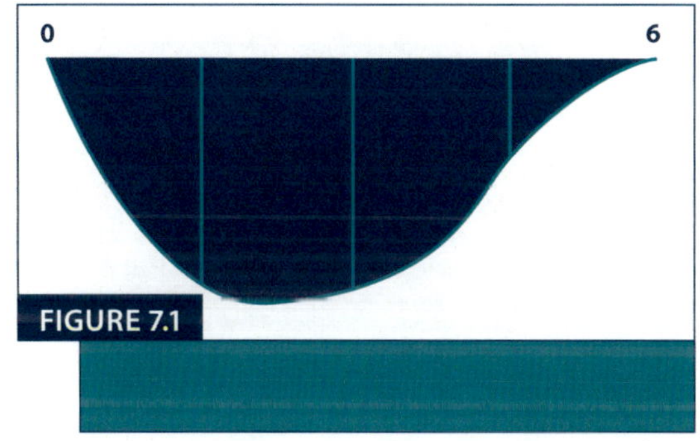

a. Your three depth measurements are equal distances apart and equal distances from the banks of the stream. How much distance is between each depth measurement?

b. The three depths you measure are 0.7 ft, 0.7 ft, and 0.3 ft. What is the cross-sectional area of the stream? Show your work.

c. It takes a ping pong ball 45 seconds to float 2 ft over this area. What is the velocity of the stream?

    d. What is the discharge of the stream, assuming it has a rocky bottom? Show your work.

    e. Another geologist is looking over your work, and he is able to infer that location X is upstream of locations 1 and 2 even though you didn't label it and he didn't see it. How could he tell?

3. You go to a location downstream of 1 and 2. It has the same cross-sectional area as location X, but a higher discharge. How is this possible?

4. You want to measure the discharge in an area where the cross-sectional area changes dramatically with distance. What is the best way to estimate cross-sectional area to determine discharge?

    a. Use the cross-sectional area of the downstream location only.

    b. Use the cross-sectional area of the upstream location only.

    c. Take the average of the upstream and downstream cross-sectional areas.

If you finish this section but you're still sitting at the picnic tables, please read the procedures in the next section.

PART 2: FIELD WORK—MEASURING DISCHARGE

Today you'll be measuring the discharge in Battle Branch Creek. Each group will be measuring discharge along a different stretch of the creek, and then we'll compare class data to data obtained from a USGS hydrograph station.

1. Which discharge sampling location were you assigned? (**Circle one.**)

   ◎ **A   B   C   D   E**

   Look on the last page of the lab to see where the site is!

2. Would you describe the sediment as sandy (sand-sized sediments) or rocky (gravel or larger sized sediments)? In addition to the one-word descriptor, include a more detailed description of what you observe, e.g., the angularity of the sediments, the mineral or rock composition, and any changes in sediment size/shape as you move across the stream.

3. How much distance will you put between your "start" and 'finish" line for testing velocity? Keep in mind: (a) you want the cross-sectional area to remain relatively constant along this distance, and (b) if the distance is too small, it will be difficult to measure velocity accuracy. If you make the distance too far, it will take forever! My suggestion: put the ping-pong ball in the water and see how far it moves in about 30–60 seconds, then make that the distance you use for your velocity calculation.

   a. What is the distance between your start and finish line in feet? _____________

   b. What is the stream width at the upstream location? Write this in Table 1, Column E. _____________

   c. If you want to take 3 depth measurements at equal distances across the stream, how much distance should there be between depth measurements? (**Hint:** See question 2) Write this value in Table 1, Column B. _____________

   d. What is the stream width at the downstream location? Write this in Table 1, Column E. _____________

   e. If you want to take 3 depth measurements at equal distances across the stream how much distance should there be between depth measurements? (**Hint:** See question 2) Write this value in Table 1, Column B. _____________

4. Before taking your depth measurements, you may want to use the tent stakes and string provided to create a line going across the width of the stream. This will act as your start and stop line for measuring velocity via ping pong ball, and it may help you stay in a straight line as you take your depth measurements.

5. Filling out Table 7.1:

   a. Fill out Column B based on your answers to #3.

   b. Fill out Column C by measuring water depths at 3 distances across your upstream (start line) and downstream (finish line) location. **Hint:** To convert inches to feet, divide by 12.

   c. Fill out Column D: Average your values from Column C to calculate an average depth value for your upstream and downstream location. **Note:** Your answer should be in feet by this time!

   d. Fill out Column E based on your answers to #5

   e. Fill out Column F by multiplying the values in Columns D and E.

Data for calculating cross-sectional area.

TABLE 7.1

A	B	C			D	E	F
Sample Location:	Distance between depth sampling locations	Water Depth 1 (ft)	Water Depth 2 (ft)	Water Depth 3 (ft)	Water Depth average (ft) *(average all C columns)*	Width of stream (ft)	Cross-sectional area (ft^2) *(column D × E)*
Station 1: upstream							
Station 2: downstream							

6. Average together the two values you calculated for Column F to get an overall cross-sectional area for your section of the stream. Write your answer in the box to the right.

Average Values.

7. Now that you have your cross-sectional area, you can get the discharge of the stream if you know the velocity of the water.

   a. Time how long it takes your ball to float from your starting location to your ending location.

   b. Do three trials and average your results. Add data to Table 7.2.

   c. Calculate average velocity by dividing the distance between your start and stop line (also your answer to 7a) by the average time.

Data for determining water velocity.

TABLE 7.2		TRIAL 1	TRIAL 2	TRIAL 3	AVERAGE
Time (sec)					
Average Velocity (ft/sec)					

8. Calculate stream discharge at your location. **Discharge = Area × Velocity × Coefficient** *Use your average cross-sectional area from #10, and your average velocity from Table 2. For the coefficient, use 0.9 if the stream bottom along that stretch is mostly sandy, use 0.8 if it is mostly rocky.*

   Note: This is essentially the same equation they use in the video → there, they use: Flow = ALC/T, where flow = discharge, A = area, C = coefficient, and L/T = length/time (velocity).

   Discharge= _______________________________ cfs (cubic feet per second, ft³/sec)

9. ***PLEASE DO THIS STEP WITHIN 2 HOURS OF YOUR LAB PERIOD SO THAT DATA IS AVAILABLE TO OTHERS.*** Compile data in the shared OneDrive document for this lab. You can access this in the "Lab 7 materials" section of Sakai. You ***must*** be signed in to your UNC account in order to access this link.

   Answer these questions once all data is available (starting Wednesday evening at 9 pm).

   a. How does your discharge compare to others who looked at the same location:

      i. How much variability is there in the data for your site?

      ii. What causes this variability? Comment on human error, weather changes, methodology differences, and other factors that could contribute to the observed error.

   b. How does your discharge compare to upstream/downstream locations:

      i. Does the discharge increase, decrease, stay constant, or vary randomly as you go downstream?

      ii. Explain the data using what you know about how discharge SHOULD vary as you move downstream, the distances between sample sites, methodology differences, human error, etc.

PART 3: HISTORICAL DATA

1. Go to this website, which shows data collected from the USGS hydrograph station located along Battle Branch Creek: *https://waterdata.usgs.gov/nwis/inventory/?site_no=0209736050&agency_cd=USGS.* **This site is on Battle Creek, *downstream* of Battle Park. How would you expect discharge data to compare to the sites you tested in lab? Justify your answer.**

When you click on "monthly statistics," then click the box to select "discharge, cubic feet per second, and click "submit," you see Table 7.3. The average discharge data for each month from October 1996 to September 2001 should be visible. Use this data to help you answer the two next questions.

00060, discharge, cubic feet per second.

TABLE 7.3

YEAR	MONTHLY MEAN IN Ft^3/S (CALCULATION PERIOD: 1996-10-01 → 2001-09-30)											
	JAN	FED	MAR	APR	MAY	JUN	JUL	AUG	SEP	OCT	NOV	DEC
1996										0.691	0.523	0.266
1997	0.262	0.419	0.343	0.486	0.262	0.843	0.554	0.134	1.00	0.545	0.471	0.278
1998	1.66	1.92	2.34	0.373	0.223	0.129	0.098	0.199	0.359	0.203	0.255	0.524
1999	1.76	0.470	0.741	0.513	0.109	0.083	0.187	0.684	9.74	0.605	0.476	0.346
2000	1.34	1.26	0.620	1.15	0.459	0.265	0.750	0.403	0.379	0.031	0.286	0.174
2001	0.252	0.325	1.19	0.445	0.294	0.842	0.120	0.419	0.095			
Mean of Monthly Discharge	1.1	0.88	1.0	0.59	0.27	0.43	0.34	0.37	2.3	0.42	0.40	0.32

**No Incomplete data have been used for statistical calculation

2. Average *all* of the group data shown on the OneDrive document for this semester's lab to get one average value. Leave off any values that are drastically different.

   a. What is the average discharge? ____

   b. How does your average compare to the monthly mean (average) values for October from 1997 to 2000? Be specific, and discuss how the downstream location of this hydrograph may affect comparisons.

3. Look at how average discharge values vary month-to-month and between Octobers in the table above.

   a. Docs seasonal variance match what you would expect based on the pre-lab videos? Describe the pattern observed, if any.

   b. Does your calculated average discharge fit into the range for October values?
      Yes   or   No

The tables below show temperature and precipitation data for the time period measured by the hydrographs (above) and all of the 2019 data available so far. Use these tables and the discharge table (previous page) to help you answer the next question. This data was collected by NOAA at the Chapel Hill 2 W Weather Station at 35.9086°, –79.0794°, from *https://www.ncdc.noaa.gov/cdo-web/datasets/GSOM/stations/GHCND:USC00311677/ detail.*

TABLE 7.4

	1997				1998				1999			
	T_{avg} (°F)	$T_{max,avg}$ (°F)	$T_{min,avg}$ (°F)	PRECIP (IN)	T_{avg} (°F)	$T_{max,avg}$ (°F)	$T_{min,avg}$ (°F)	PRECIP (IN)	T_{avg} (°F)	$T_{max,avg}$ (°F)	$T_{min,avg}$ (°F)	PRECIP (IN)
Jan	40.0	50.8	29.2	3.78	43.4	52.6	34.1	8.93	44.5	57.0	32.0	7.62
Feb	45.7	56.5	34.8	2.89	44.0	55.1	32.9	6.60	43.9	56.7	31.1	2.33
Mar	53.8	65.9	41.7	2.45	49.3	60.1	38.6	7.64	46.8	60.1	33.4	3.54
Apr	55.1	66.7	43.6	6.20	58.6	70.6	46.6	3.18	60.0	72.7	47.2	4.62
May	64.2	76.6	51.9	1.37	68.8	79.4	58.1	3.07	65.7	77.7	53.7	1.10
Jun	71.1	81.1	60.8	3.12	76.2	87.1	65.3	2.46	72.9	82.7	63.0	2.93
Jul	79.1	89.4	68.9	5.80	79.8	91.0	68.5	2.24	80.1	89.7	70.4	2.87
Aug	76.4	87.6	65.1	0.79	78.2	89.5	66.8	4.30	79.8	90.6	68.9	5.45
Sep	71.3	82.3	60.3	4.07	74.5	86.0	62.9	3.93	69.1	79.1	59.0	24.01
Oct	59.8	71.9	47.6	3.48	61.3	74.9	47.6	1.72	57.8	69.1	46.5	2.54
Nov	46.8	57.8	35.8	4.61	51.0	64.2	37.7	2.83	54.9	68.4	41.3	2.52
Dec					46.4	56.8	36.0	4.04	42.8	54.9	30.8	1.74

	2000				...	2019			
	T_{avg} (°F)	$T_{max,avg}$ (°F)	$T_{min,avg}$ (°F)	PRECIP (IN)		T_{avg} (°F)	$T_{max,avg}$ (°F)	$T_{min,avg}$ (°F)	PRECIP (IN)
Jan	37.8	49.2	26.4	4.61		41.5	52.0	31.1	3.78
Feb	44.5	57.6	31.4	2.25		45.1	56.0	34.3	5.22
Mar	53.4	66.6	40.3	2.68					
Apr	56.7	68.2	45.1	6.25		60.0	70.8	49.1	4.78
May	69.6	82.4	56.8	2.53		73.1	84.1	62.1	1.99
Jun	76.6	87.1	66.0	3.19		75.5	86.4	65.7	5.18
Jul	76.0	85.0	67.0	7.69		81.2	92.3	70.1	3.95
Aug	75.3	85.3	65.2	5.0		77.6	88.1	67.2	3.92
Sep	69.3	78.8	59.8	4.22		76.4	88.0	64.7	0.51
Oct	60.1	74.6	45.6	0.26					
Nov	47.2	59.6	34.9	2.37					
Dec									

4. Does precipitation influence discharge? Justify your answer using specific examples comparing data from both the climate and discharge tables.

5. How long (in terms of months) do large precipitation events impact discharge in Battle Branch Creek? Look at the September 1999 precipitation event as evidence for your claim.

6. Based on the average discharge calculated for this lab, do you think October 2019 was a high- or low- rainfall month compared to the average? Justify your answer.

FIGURE 7.2

Battle Park Pedestrian Trails
Boundary St.
E. Franklin St.
Park Place
Glendale Dr.
Glenhill Ln.
Park Place Parking Lot
Monarch of the Forest
OWASA Access (0.7)
OWASA Access (0.7)
Battle Branch Creek
Battle Branch Creek
Solitary Hill Trail (0.1)
Historic Spring
Cedar Ridge
Cedar Top Trail (0.2)
Bent Beech Trail (0.3)
Copperhead Curve Trail
To Community Center →
Location D
Location B
Location A: Creek Stop 1
Location C: Creek Stop 2
Location E
Old Poplar Picnic Place
Lovers' Leap Lookout
Sourwood Loop (0.4)
Deer Track Trail (0.2)
Rainy Day Trail (0.3)
University Trail (0.1)
Forest Theatre
Meet here
Jackson Hall
Cobb Dorm
Country Club Rd.
Center for Dramatic Art
Sisters' Corner
Gimghoul Rd.
Glandon Dr.
Ridge Ln.
Evergreen Ln.
Lovers' Stone Chair
Gimghoul Castle
Piney Prospect Overlook
Lovers' Loop (0.4)
N
0   0.05 Miles
Contour Interval: 2 Feet
Trail length is given in miles
Picnic Area
Bench
Kiosk

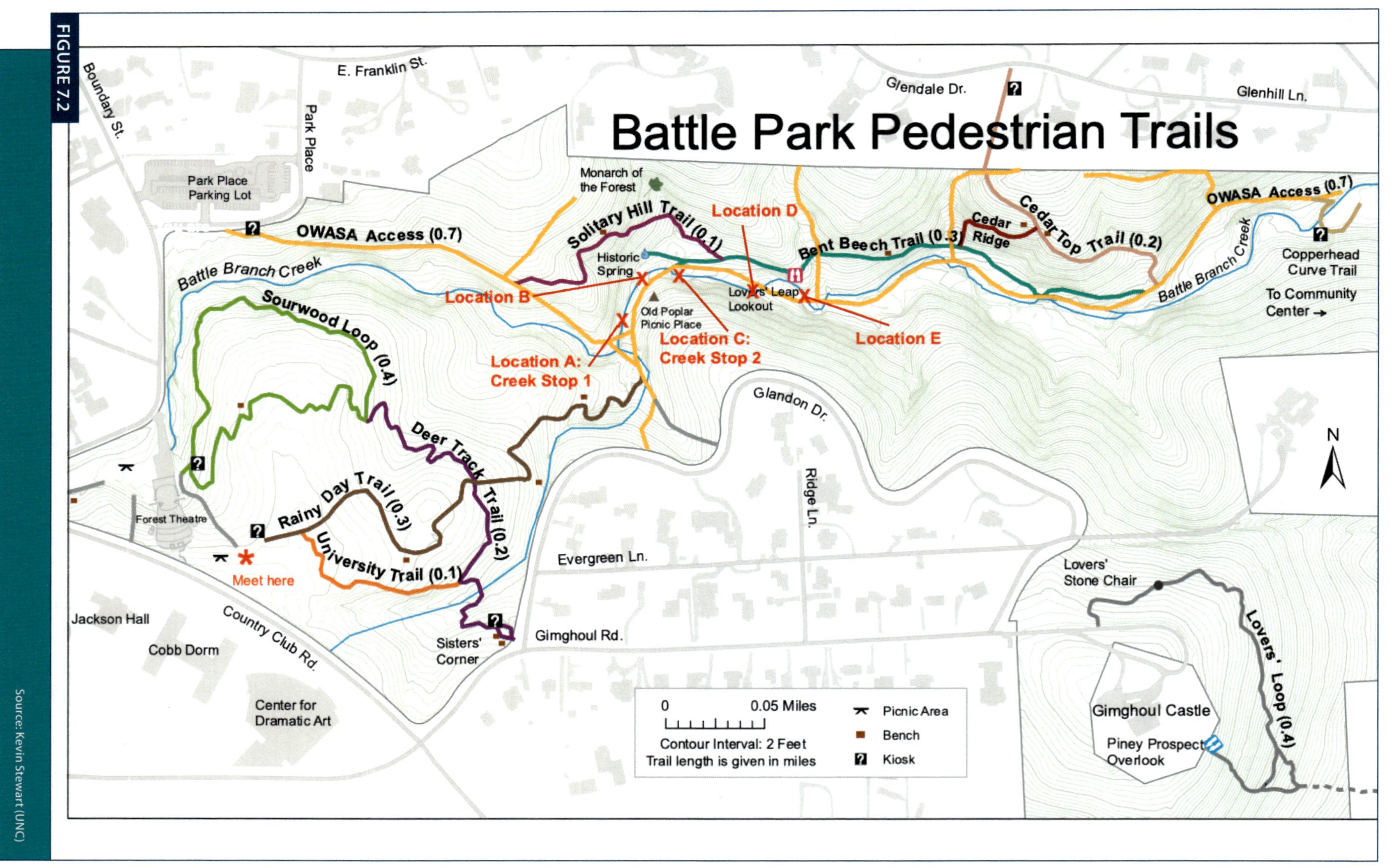

EARTH'S INTERIOR STRUCTURE

Name: _________________________ Section: __________ Date: _________

*This lab is adapted from IRIS's lab found at: https://www.iris.edu/hq/inclass/lesson/
determining_and_measuring_earths_layered_interior*

PREFACE

INTRODUCTION

A dialogue you overhear …

Kristin: I don't understand how we know the Earth's core is iron. How can we possibly know what the Earth's interior is made of? We've never seen anything deeper than the crust! It could just be the same rocks we see on the surface all the way down.

Luis: Well, we know seismic waves change speed as they move through materials of different densities and physical states—whether the material is solid or liquid. Seismic waves change in speed as they go through the Earth; that means the Earth's interior isn't all the same. We can tell if the Earth's interior is changing in density and physical state, and we can even see the depth at which each layer begins based on change in seismic wave speed.

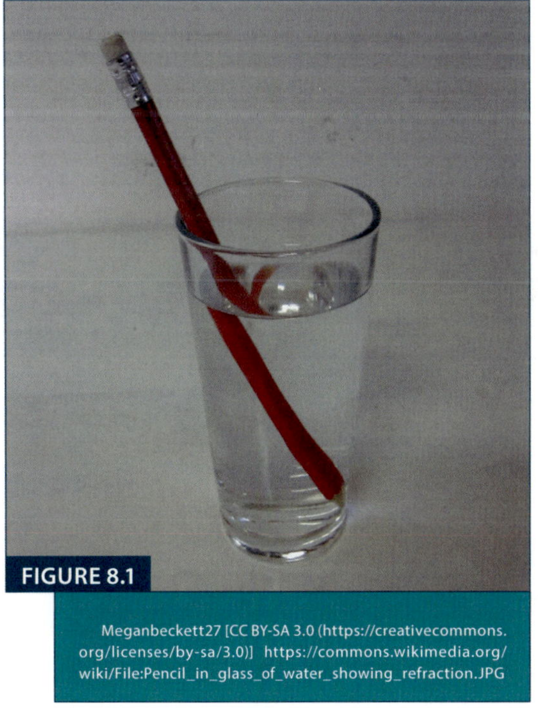

FIGURE 8.1

Meganbeckett27 [CC BY-SA 3.0 (https://creativecommons.org/licenses/by-sa/3.0)] https://commons.wikimedia.org/wiki/File:Pencil_in_glass_of_water_showing_refraction.JPG

Kristin: Well, how do you know that the seismic waves change speed?

Luis: Look at this pencil in a glass of water. Light acts as a wave, and light waves move at different speeds in water, air, and glass. The bending of the light where it reaches a material of a different density makes the pencil appear broken. Seismic waves do the same thing—they change speeds when they hit a material with a different density!

Kristin: Ok… but we can SEE the light bending. How can you see where seismic waves bend? How do you know when they start travelling at a different speed?

Luis: …

Today, we're giving you the supplies you need to help resolve Kristin's question:

⊙ **How can we determine the depth at which seismic waves bend, and determine if they've changed speed as they travel through the Earth?**

LEARNING OBJECTIVES

Students will be able to …

⊙ Read a seismogram to determine p- and s-wave arrival time.

⊙ Create a scale model of Earth's interior layers.

⊙ Determine the depth at which the Earth's density changes based on changes in seismic wave speed and the direction of density change (increase or decrease) at observed boundaries.

⊙ Explain how models are used in combination with observational data to discover phenomenon that are not directly observable.

MATERIALS

⊙ Please print out this lab, including all appendices, and bring to class.

⊙ Graph paper

⊙ Ruler, protractor, and a compass

BEFORE YOU BEGIN

1. From your pre-lab, describe how you would expect wave speed to change with density: (**Hint:** To answer this question, you'll have to combine information from multiple pre-lab videos.)

   a. If density *increases,* wave speed will ______________________________

   b. If density *decreases,* wave speed will ______________________________

2. (**Circle the best answer.**) When seismic waves leave the site of an earthquake:

   a. Waves move away from the focus on the Earth's surface only.

   b. Waves move away from the focus through the Earth's interior and along the surface.

   c. Waves move away from the focus into the Earth's interior only.

3. (**Circle the best answer.**) Which wave arrives at a seismic station first?

   a. *P*-wave   b. *S*-wave   c. Surface wave

4. (**Circle the best answer.**) Which statement is *false?*

   a. The further a seismic station is from the site of the earthquake, the deeper inside the Earth *p*- and *s*- waves travel.

   b. The further a seismic station is from the site of the earthquake, the longer it takes *p*- and *s*- waves to get there.

   c. The further a seismic station is from the site of the earthquake, the more likely it is that waves travelling along the surface will arrive before waves travelling inside the Earth.

5. Make some predictions about what would happen in the following scenarios. Your prediction should relate to velocity, (e.g. increase in speed, decrease in speed, constant speed), and to the magnitude or speed of the change, (e.g., increases/decreases significantly, increases/decreases suddenly).

   a. Predict how the speed of seismic waves would vary if Earth's interior does not change.

   b. Predict how the speed of seismic waves would vary if Earth's interior has gradual density differences, e.g. a gradual increase in density.

   c. Predict how the speed of seismic waves would vary if Earth's interior has sudden density differences at specific locations, e.g. a sudden increase in density.

PART 1: ANALYZING OBSERVED DATA: DOES *P*-WAVE VELOCITY CHANGE BASED ON DISTANCE *P*-WAVES TRAVEL?

PROCEDURE

The first step in determining whether the speed of seismic waves changes as you travel through the Earth is to see how quickly waves travel a measured distance.

DATA COLLECTION

Tl;dr: Record *p*-wave arrival time. Compare to distance each station is from earthquake epicenter and use velocity = distance/time and the conversion 1 geocentric degree = 111 km to figure out *p*-wave velocity.

More detailed instructions:

1. Open Excel file available on Sakai. Look at print-out or Sakai version of seismograms (**Figure 1**).

2. Determine the distance each seismograph station is from the epicenter of the earthquake. Distances are reported in geocentric degrees. Record data on Excel **Table 1** Column B.

3. Read the *p*-wave arrival time for each seismogram. Record data on Excel **Table 1,** Column C. **Note:** There are 60 seconds in a minute. 30 seconds should be recorded as 0.5 mins, etc.

4. Convert the arrival time of the *p*-waves to seconds by multiplying time by 60. This will make velocities easier to compare. Record data on Excel **Table 1,** Column D.

5. Convert geocentric degrees to km by multiplying by 111. Then, the distance through the Earth (in km) the *p*-wave travels will be calculated automatically in the spreadsheet. Record values in **Table 1,** Column E.

6. Calculate the velocity of the *p*-waves in km/sec (velocity = distance/time) in **Table 1,** Column F. **Hint:** Don't do calculations individually! Instead write a formula in Excel—your TA will demonstrate how to do this.

DATA ANALYSIS

7. Is the *p*-wave velocity constant always the same? (Yes / No) **(Circle one.)**

   a. If yes, write the *p*-wave velocity here: _______________________________

   b. If no, write the *average* *p*-wave velocity here: _______________________, and use the space below to describe the relationship between *p*-wave velocity and distance, if any.

8. **(Circle the best answer.)** Assuming the shortest linear path inside the Earth is the way the fastest *p*-waves are travelling to each location, how does the maximum depth in the Earth compare to distance travelled from the epicenter? (**Hint:** *Look at the diagram to the right to help you with this question; the earthquake epicenter is at the top, and the seismic stations are triangles.*)

   a. *P*-waves travel at deeper depths in the Earth the further they move from the earthquake site.

   b. *P*-waves travel at shallower depths in the Earth the further they move from the earthquake site.

   c. Depth of *p*-waves is not related to distance they travel away from the earthquake site.

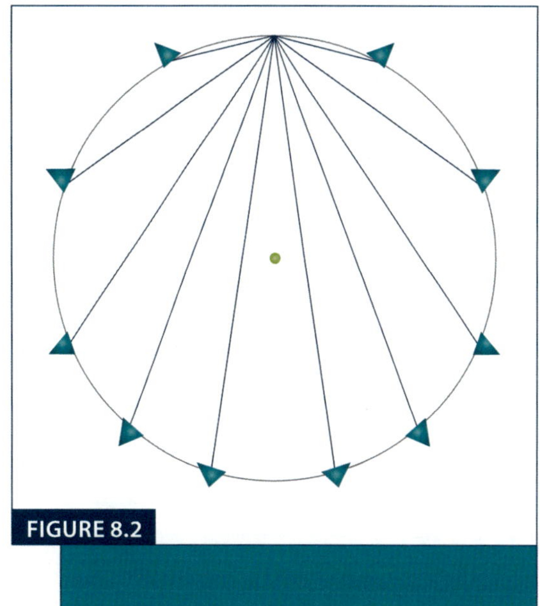

FIGURE 8.2

9. **(Circle the best answer.)** How does the depth of travel compare to the speed of the *p*-waves?

   a. The deeper *p* waves travel, the faster they go.

   b. The deeper *p*-waves travel, the slower they go.

   c. The depth of *p*-waves is not related to velocity.

CONCLUSIONS FROM PART 1

1. Based on your data and analysis, describe how Earth's interior changes with depth. Justify your answer.

2. Are the data you've collected and analyzed sufficient to answer this question: How can we determine the depth at which seismic waves bend, and determine if they've changed speed as they travel through the Earth? If not, suggest what else you may need to measure.

PART 2: ANALYZING MODELED DATA IN CONJUNCTION WITH OBSERVED DATA—HOW CAN WE DETERMINE THE DEPTH AT WHICH THE VELOCITY CHANGE OCCURS?

PROCEDURE

A good method for confirming or testing your answer from Part 1 is to generate a model. This model will show you what the data ***should*** look like if the conditions you specify are true. You're going to specify that the velocity is a constant, and then compare data you collect making this assumption to data from real-world observations.

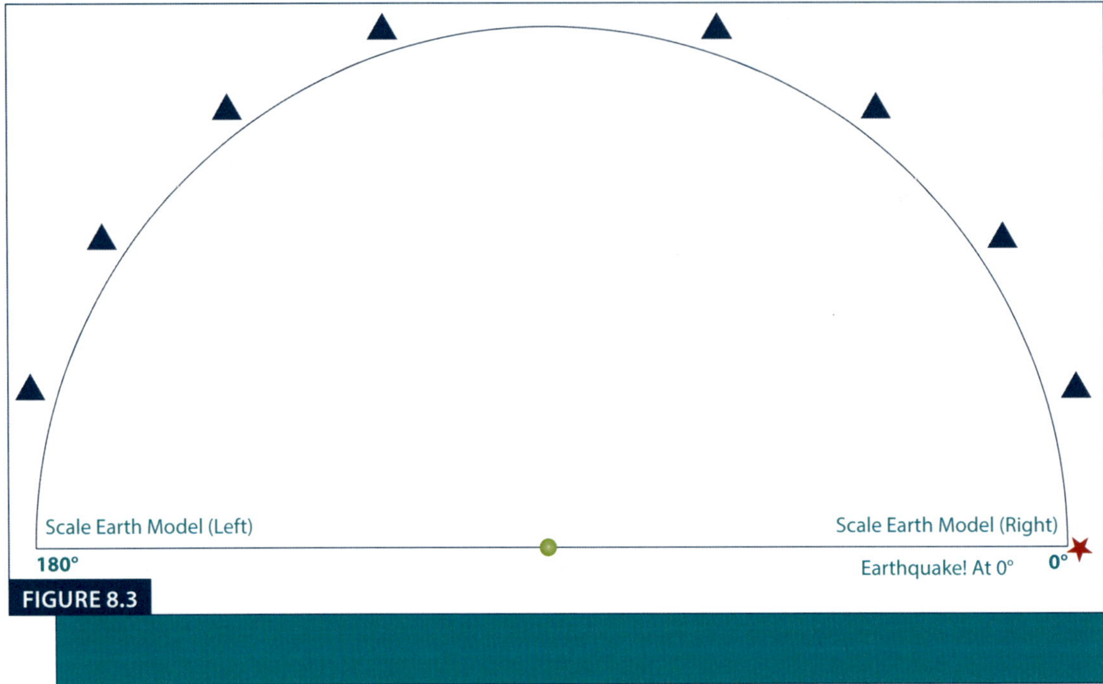

FIGURE 8.3

DATA COLLECTION

Tl;dr: On the model of the Earth provided, draw a line connecting the earthquake site to each seismic station. Measure the line, convert to Earth distances, and use your average velocity to Part 1 to calculate the time at which *p*-waves should arrive at each station if *p*-waves move at a constant velocity.

Detailed instructions on next page.

DETAILED INSTRUCTIONS

1. Go to the next tab of the Excel file and find Excel **Table 2.**

2. Use Figure 8.3, the diagram of Earth provided and draw a line connecting the earthquake site to each seismic station.

3. Measure the lines using a ruler. Write the length of each line in **Table 2,** Column C.

4. Convert the ruler/model distance to a distance on Earth by multiplying the length of the line by the conversion factor, and record this value in **Table 2,** Column D.

   Hint: Don't do calculations individually! Instead write a formula in Excel—your TA will demonstrate how to do this.

   a. To find the conversion factor:

      i. Measure the radius of the diagram of Earth in cm: ________________

      ii. Use this equation to determine what 1 cm on the diagram represents in km on Earth. Plug your value from 4.a.i. in for the "Radius of Earth" and solve for X.

$$\frac{1 \text{ cm }_{(diagram)}}{X \text{ km }_{(Earth)}} = \frac{\text{Radius of Earth (cm)}_{(diagram)}}{6371 \text{ km }_{(Earth\ radius)}}$$

      iii. Conversion factor (x): ________________

5. Use your answer from Part 1, #6 to determine the time *p*-wave should arrive at each station assuming an average/constant velocity. Write the "modeled travel time" in **Table 2,** Column E.

6. Convert the "modeled travel time" to minutes by dividing by 60. Record this in **Table 2,** Column F. This will allow you to compare modeled data to observed data more easily.

DATA ANALYSIS

1. Visualize your modeled and real seismic data in a graph. The graph should be generated automatically on the 3rd sheet of the Excel file if you've filled out the tables. Describe what this graph is showing you, e.g., what kind of data are being compared, and the types of conclusions that can be drawn from this data.

2. **(Circle the best answer.)** Which statements best describe the relationship between the two data sets?

    a. At all distances (0°–180°), the two datasets match well.

    b. At all distances (0°–180°), the two datasets **do not** match well.

    c. When distances are short (less than 100°), the two datasets match well.

    d. When distances are long (more than 100°), the two datasets match well.

3. **(Circle the best answer.)** Between 100° and 120°, there is a sudden shift in the observed/actual data. This sudden shift shows:

    a. Actual p-waves are now moving faster than the model predicts they should

    b. Actual p-waves are now moving slower than the model predicts they should

4. Due to the sudden shift, we can infer that the deviation in actual and modeled p-wave arrival time between 100 and 120° from the earthquake focus occurs because seismic waves that travel this far go deep enough into the Earth to hit another layer of the Earth.

    ◉ Does this layer make p-waves   speed up   or   slow down? **(Circle one.)**

CONCLUSIONS FROM PART 2

1. Based on your data and analysis, describe how Earth's interior changes with depth. Incorporate information from Parts 1 and 2 to justify your answer.

2. Are the data you've collected and analyzed sufficient to answer the following question: How can we determine the depth at which seismic waves bend and determine if they've changed speed as they travel through the Earth? If not, suggest what else you may need to measure.

PART 3: HOW CAN WE USE DISTANCE FROM THE FOCUS TO DETERMINE THE DEPTH IN THE EARTH WHERE WE MOVE FROM 1 LAYER TO ANOTHER?

PROCEDURE

We will use a scale diagram of the Earth (Figure 8.4) and the graph you generated in Part 2 to figure out the depth at which we transition from one layer to another.

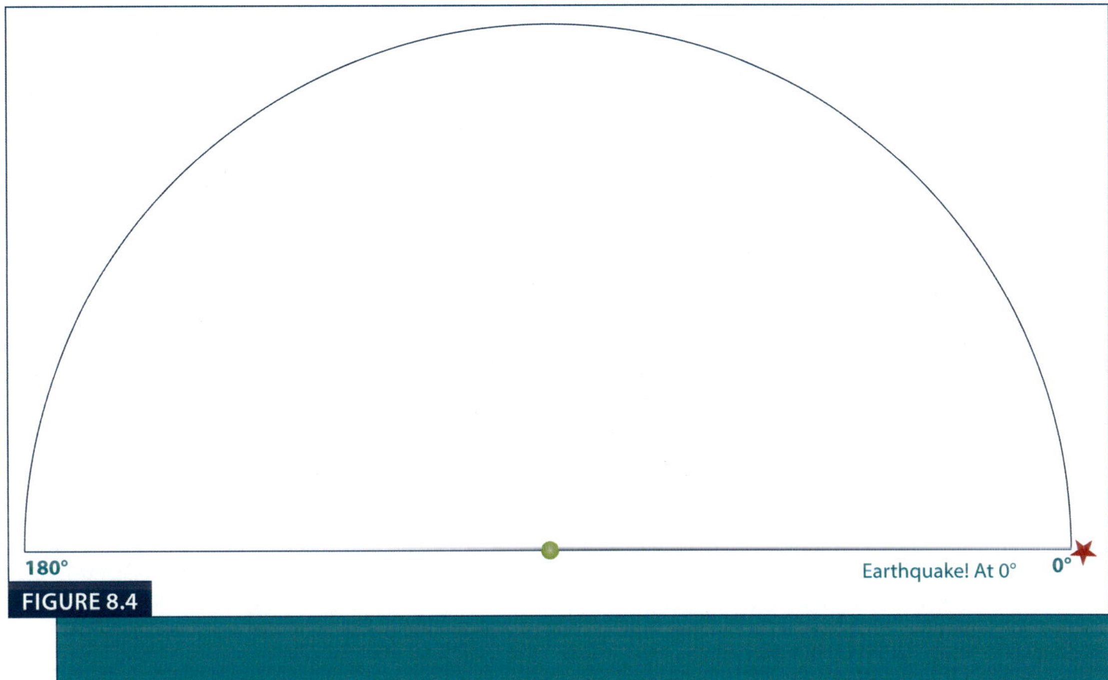

FIGURE 8.4

DATA COLLECTION

Tl;dr: On the diagram of the Earth provided, draw a line connecting the earthquake site to seismic stations at the distance you saw modeled and observational data diverging (between 100 and 120 geocentric degrees). Determine how close seismic waves get to the center of Earth.

DETAILED INSTRUCTIONS

1. Look at the graph you generated in Part 2, and decide the distance from the epicenter (in geocentric degrees) at which the velocity transition occurred (**Hint:** Between 100° and 120°)

    Distance at which velocity transition occurs: ________________________ °

2. Using Figure 8.4, the diagram of Earth provided (previous page):

    a. Place your protractor so that it is centered in the center of the Earth.

    b. Making the location of the earthquake 0°, mark the angle from #1 (example for where 140° would be is given on the right in Figure 8.5).

    c. Draw a straight line from the epicenter to the spot on the Earth's surface that corresponds to the geocentric angle you measured.

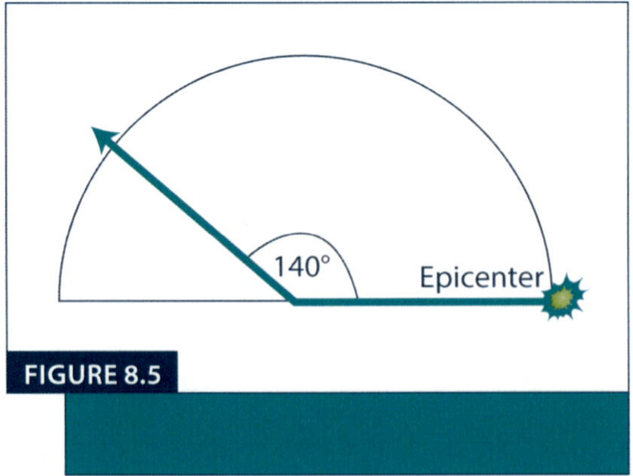

FIGURE 8.5

3. Find the spot along this line where it is closest to the Earth's center. This is the maximum depth the *p*-wave that travels a straight-line path from the earthquake focus to a seismic station at that distance would travel. You may want to use a ruler or compass to do this.

    Maximum depth *p*-wave travels (distance from Earth's center): ______________ (cm)

4. Convert the ruler/model distance to a depth on Earth by multiplying the length of the line by a conversion factor which you can determine using the formula below.

    a. To find the conversion factor:

        i.   Measure the radius of the diagram of Earth in cm: ______________

        ii. Use this equation to determine what 1 cm on the diagram represents in km on Earth. Plug your value from 4.a.i. in for the "Radius of Earth" and solve for X.

$$\frac{1\text{ cm }_{(diagram)}}{X\text{ km }_{(Earth)}} = \frac{\text{Radius of Earth (cm) }_{(diagram)}}{6371\text{ km }_{(Earth\ radius)}}$$

        Conversion factor (x): ______________________

    Maximum depth *p*-wave travels (distance from Earth's center): ______________
    (km on Earth)

DATA ANALYSIS

1. **(Circle the best answer.)** What happens to the depth seismic waves travel within the Earth as they move further from the earthquake focus? (**Hint:** You may have to draw more lines to figure this out.)

    a. Maximum depth increases with distance

    b. Maximum depth decreases with distance

    c. Maximum depth varies, but does not correlate with distance

    d. Maximum depth is constant with distance

2. Compare the depth you calculated to the layers shown in the diagram on the right. *Note:* Figure 8.6 shows distances in km between each boundary, not total distance from Earth's center!

    a. Which boundary did you identify in Part 2?

    b. When seismic waves cross this boundary, do they speed up or slow down?

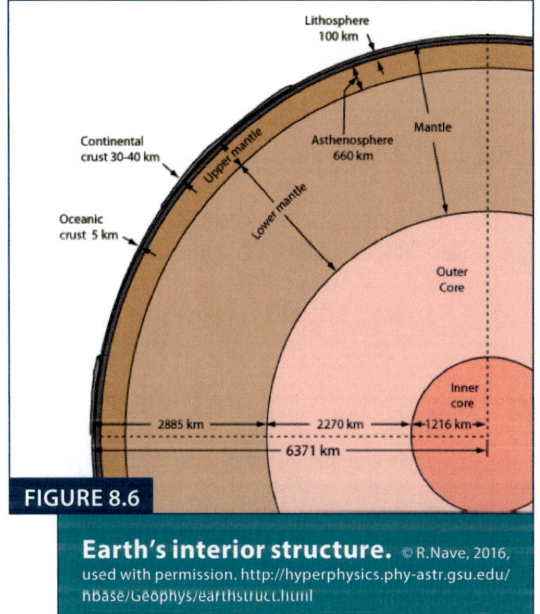

FIGURE 8.6

Earth's interior structure. © R.Nave, 2016, used with permission. http://hyperphysics.phy-astr.gsu.edu/hbase/Geophys/earthstruct.html

3. Look at the graph you completed in Part 1.

    a. Where else do you see a potential transition (express as a distance in geocentric degrees)?

    b. Describe the procedure you could use to determine if that distance in geocentric degrees translates to a depth in the Earth corresponding to one of the layers shown.

CONCLUSION

1. Are the data you've collected and analyzed sufficient to answer the following question: How can we determine the depth at which seismic waves bend and determine if they've changed speed as they travel through the Earth? If not, suggest what else you may need to measure.

FIGURE 8.7

11/18/2019
Earthquake: M7 Haiti Region
Magnitude 7, Haiti Region, 12 Jan 2010 @ 21:53UTC
The above earthquake recorded at 14 seismic stations:
Time Since the Earthquake (hour:min:sec)
Seismograms by Distance from the Earthquake (1 degree = ~110 km)
ARU Arti, Russia
BJT Baijiatuan, Beijing, China
ENH Enshi, Hubei Province, China
KAPI Kappang, Sulawesi, Indonesia
KDAK Kodiak Island, Alaska, USA
KIEV Kiev, Ukraine
KURK Kurchatov, Kazakhstan
PALK Pallekele, Sri Lanka
PTGA Pitinga, Brazil
SACV Santiago Island, Cape Verde
SDCO Great Sand Dunes National Park, Colorado, USA
SDV Santo Domingo, Venezuela
TRQA Tornquist, Argentina
WRAB Tennant Creek, NT, Australia

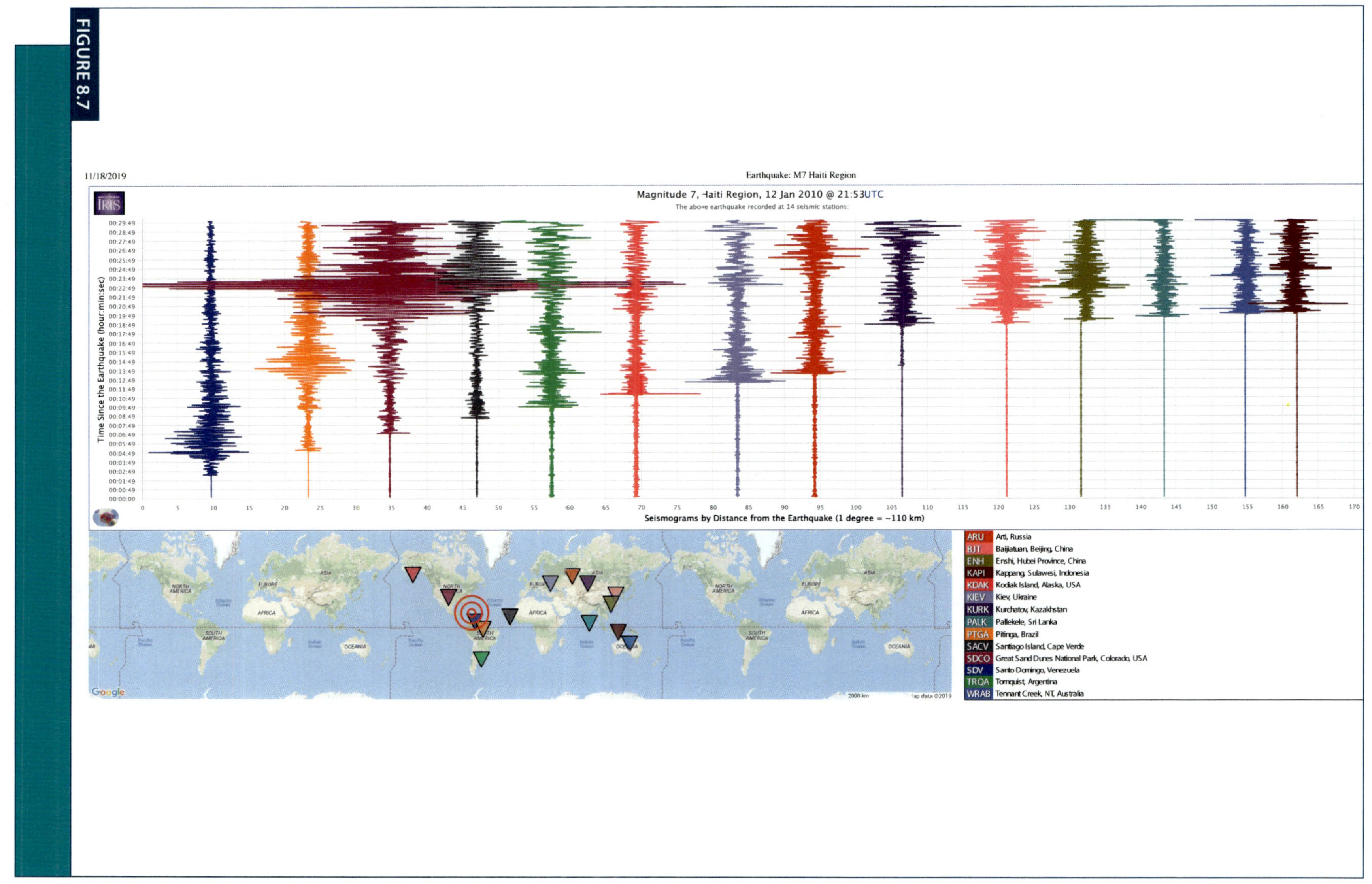

EARTHQUAKES

Name: _________________________ Section: __________ Date: _________

Part 1 adapted from Ben Surpless' Earthquake Lab. Part 2 adapted from IRIS's Exploring Rates of Earthquake Occurrence:
https://www.iris.edu/hq/inclass/lesson/exploring_rates_of_earthquake_occurrence

PREFACE

INTRODUCTION

A dialogue you overhear …

Kristin: Ok, so we looked at seismic waves last week. And I know they come from earthquakes, but how do earthquakes create waves?

Luis: Seismic waves are just waves of energy that are released when the earth moves.

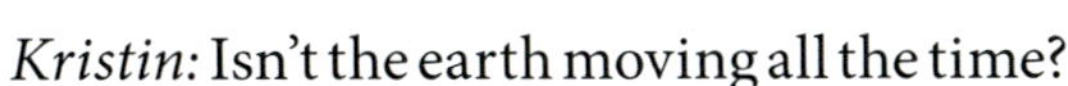

Kristin: Isn't the earth moving all the time?

Luis: You're right … the ductile *mantle* is moving all the time. But the brittle lithosphere can't move quite as easily—instead, as the mantle convects, it puts stress on the lithosphere. The lithosphere elastically deforms due to the stress until it finally snaps and breaks.

Kristin: So rocks are elastic? And then the break.

Luis: Yes—imagine stretching a rubber band. If you stretch it far enough, it will break, snap back to its original shape, and release energy!

Kristin: Ok, so rocks stretch or something, then break and go back to their original shape and that releases the seismic waves. This seems like a strange analogy to me—rocks aren't stretchy.

Luis: Yep! That's how it works. Rocks may not seem stretchy, but they can and do act elastically. That's how they end up storing enough energy to cause high-magnitude earthquakes. And this theory is consistent with the idea that small earthquakes happen more frequently than big earthquakes.

Kristin: I'd be impressed if you could actual predict something like earthquake frequency using a rubber band as a model. Could you make a model using a rubber band that realistically shows how often earthquakes of different magnitudes occur in real life?

Today, we're giving you the supplies you need to help resolve Kristin's question:

- **Could you make a model using a rubber band that realistically shows how often earthquakes of different magnitudes occur in real life?**

LEARNING OBJECTIVES

Students will be able to …

- Plot (N)/year data for earthquake events of different magnitudes to determine the annual percent probability of future earthquake events.
- Compare modeled and observed data sets to determine the applicability of a simplified model to a real-world dataset.
- Compare earthquake data from multiple regions on earth to determine if earthquake behavior is the same along different fault systems.
- Quantitatively explain how the probability of earthquake events varies with magnitude.

MATERIALS

- This lab hand-out, printed, and your computer
- Physical fault model (in the lab)

BEFORE YOU BEGIN

BACKGROUND

This lab consists of 2 parts: The first allows you to simulate earthquake events so that you can see how earthquake probability is calculated, and in the second part you will look at current earthquake data to see how applicable this model is.

PART 1: CALCULATING EARTHQUAKE PROBABILITIES WITH A PHYSICAL MODEL

PROCEDURE

You will use the physical fault model to create earthquake events. The physical model replicates *elastic rebound*. Turning the crank first stretches the rubber band attached to the block. Once the stress on the rubber band exceeds the frictional force between the block and sandpaper, the block moves forward and the stress on the rubber band releases, causing it to revert back to its original shape. After you collect data, graph the data to make predictions about future earthquake events.

1. Identify how the model replicates fault behavior by drawing lines on the diagram above to match the behavior of the physical earthquake model on the left with the corresponding behaviors of a real fault system on the right.

TABLE 9.1

MODEL BEHAVIOR		FAULT BEHAVIOR
Turning the crank		Storing of elastic potential energy in rocks
Friction between sandpaper and wooden block		Fault slip that creates an earthquake
Stretching of the rubber band		Continuous, slow plate motions
Sudden slip of block		Contact between the sides of the fault

DATA COLLECTION

Tl;dr: Turn the crank (¼ turn equals 1 year) to simulate time passing. When the block moves, stop turning the crank. Record the magnitude (# of cm the block moves) and the time in Table 9.1.

More detailed instructions:

1. Start with the block at the 0 cm and the string taut (w/o stress on the rubber band).
2. One person in your group should turn the crank—each ¼ turn equals 1 year. Turn the crank slowly and steadily, counting quarter cranks, until the block moves!

3. When the block moves, record it as an event in Table 9.2 (below):

   a. Time since last event = number of ¼ turns of the crank

   b. Block position (in cm) = read off meter stick

   c. Magnitude = total change in position in cm (current-previous block position)

4. Repeat until one of the following conditions are met:

   a. You have filled out the entire table (20 total events), OR

   b. You have moved your block from start to finish twice!

      i. If you reset your block to zero, don't worry about your block position #s looking weird. Remember it's the magnitude that you need!

Modeled earthquake data.

TABLE 9.2

EVENT NUMBER	TIME SINCE LAST EVENT (YEARS)	BLOCK POSITION (CM)	MAGNITUDE (CM)
0 (start)	-----------	0	-----------
1			
2			
3			
4			
5			
6			
7			
8			
9			
10			
11			
12			
13			
14			
15			
16			
17			
18			
19			
20			

DATA ANALYSIS

1. **(Circle the best answer.)** Which of the following statements best describes the data you collected?

    a. There are many large events and few small events.

    b. The number of large and small events is relatively equal.

    c. There are many small events and few large events.

2. Fill out Table 9.3:

    a. In the "N" column, record total # of events with that magnitude.

    b. What is the total # of years covered in your modeling experiment? _____________

    c. In the "N/year" column, divide "N" by your answer to 2b.

Binned earthquake events.

TABLE 9.3

MAGNITUDE (cm)	(N)	(N)/YEAR
≤ 3		
3.1–6.0		
6.1–9		
> 9		

3. Each group will write the number of earthquakes at each magnitude and the total number of years they covered with their model on the board.

    a. Fill out Table 9.4 using class data. (the # years you use for the third column should be the total # of years used for all groups)

Binned class data.

TABLE 9.4

MAGNITUDE (cm)	(N)	(N)/YEAR
≤ 3		
3.1–6.0		
6.1–9		
> 9		

4. Describe the differences between your group's data and the class data (Tables 9.3 and 9.4) and speculate as to why you see these differences.

5. Plot your values from this table on the graph on the next page.

   a. Before plotting your points, **label the minor *y*-axis lines**. This axis is logarithmic!

   b. Plot one point for each "bin" at the midpoint of the range in magnitudes. Use Table 9.4 data. Midpoint values are 1.5, 4.5, 7.5, and 11.5. ***Note:*** If N/year for 11.5 is zero, don't plot it!

   c. Draw a straight line that goes through most of your data (if possible).

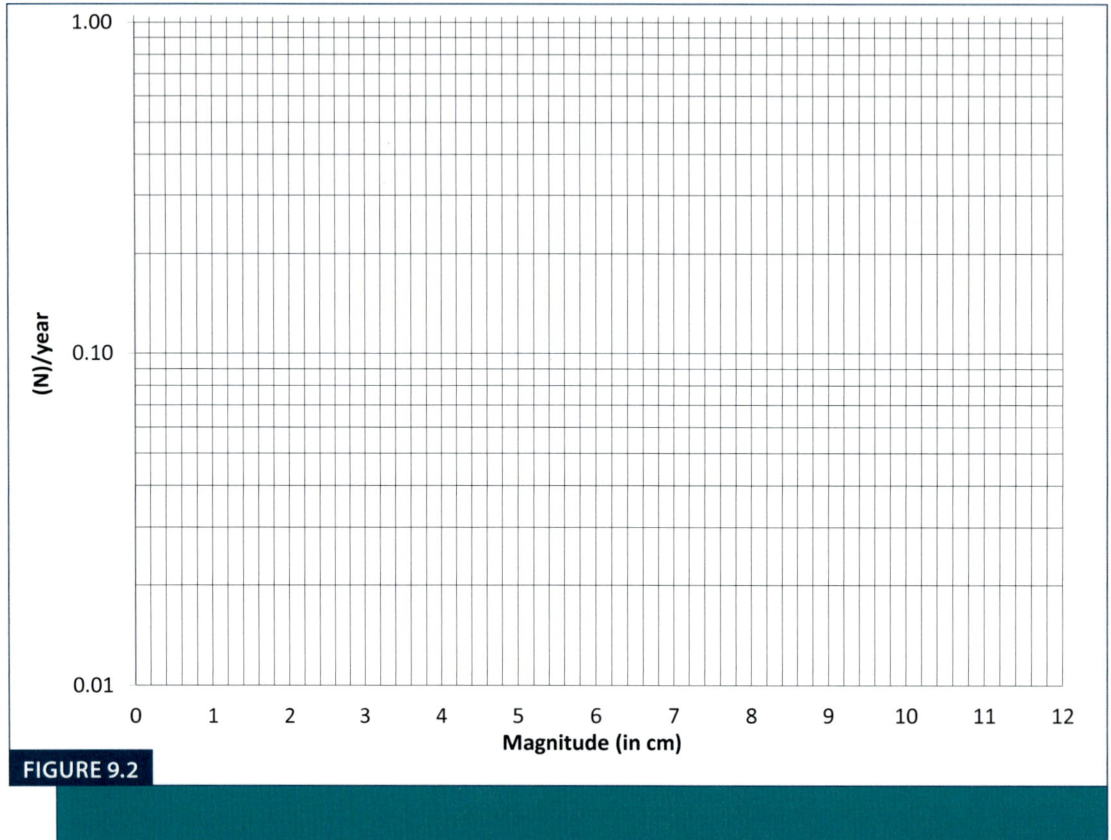

FIGURE 9.2

6. Based on your graph, what is the annual percent probability of a magnitude 7 quake in our modeled fault system? The N/year is the number of earthquakes per 1 year. Multiply that number by 100 to calculate percent probability.

7. Is this a realistic probability for real earthquakes? To test this, compare your model to models generated by the USGS for California earthquakes. Using the USGS graph, estimate the annual percent probability of a magnitude 7 quake for the California region. The N/year is the number of earthquakes per 1 year. Multiply that number by 100 to calculate percent.

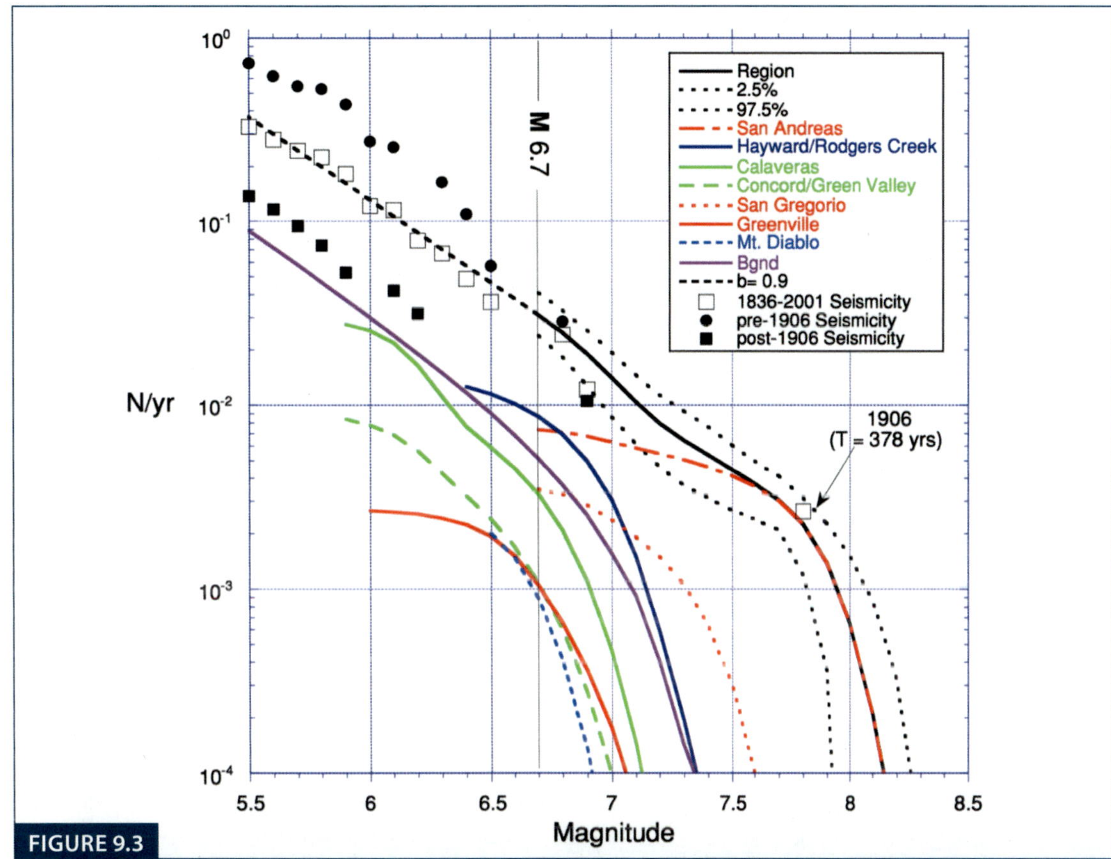

FIGURE 9.3

This is a long-term cumulative magnitude-frequency distribution for the SFBR earthquake model (USGS). The open squares, black circles, and black squares show historical seismicity data (and illustrate the significance of the 1906 earthquake on seismicity in the region). The black line illustrates the region as a whole, and each colored line refers to a specific fault. (Earthquake Probabilities in the San Francisco Bay Region: 2002-2031. Open-File Report 03-214 by Working Group On California Earthquake Probabilities 2003.) Use the dotted-to-black line for the California region. Public Domain. www.usgs.gov

8. Do you think class data and modeled California data would match better or worse for a larger earthquake, e.g., a magnitude 8. Why? Justify your answer.

PART 2: FORECASTING EARTHQUAKES USING THE IRIS EARTHQUAKE BROWSER

PROCEDURE

1. Navigate to *www.iris.edu/ieb*.
2. Manipulate earthquake magnitude on the sidebar to count earthquakes of different magnitudes over a 30-year period.

DATA COLLECTION

Tl;dr: Choose an area on www.iris.edu/ieb and count earthquake events at different magnitudes within a 30-year time period. Record the location. Do this twice (or share data with another group).

More detailed instructions:

1. Open *www.iris.edu/ieb*.
2. Zoom in to an area you're interested in looking at by double clicking on the map.
3. Click on "select new region" on the left side of the map to select a specific area (lat/long).
   a. Where in the world are you? (example answer: Japan; Texas; Middle East)

   b. What is the latitude & longitude range you selected? (shown bottom right)

4. Click on "time range" in the right sidebar and choose a 30-year period in which to collect data.
   a. Starting date: _______________________
   b. Ending date: _______________________

5. Click on the "magnitude range" on the right sidebar and choose a minimum magnitude of 8.5. Click "apply," and record the number of earthquake events (N) in Table 9.5.

6. Decrease your minimum magnitude in 0.5 degree intervals until you finish filling out the table. You never have to change the maximum magnitude as you're recording earthquake events greater than or equal to the magnitude you're plugging in.

7. **Repeat** Steps 3–6 for a second location, or team up with another group and share data so that you each have 2 locations to compare.

   a. Where in the world are you? (Location 2)

   b. What is the latitude and longitude range you selected? (Location 2)

   c. Starting and ending date (Location 2): _______________________________

8. Record data in Table 9.5.

30-Year earthquake records.

TABLE 9.5

	LOCATION 1			LOCATION 2	
MAGNITUDE	(N)	(N)/YEAR	MAGNITUDE	(N)	(N)/YEAR
≥ 8					
≥ 7.5					
≥ 7					
≥ 6.5					
≥ 6					
≥ 5.5					
≥ 5					
≥ 4.5					
≥ 4					
≥ 3.5					

DATA ANALYSIS

1. Calculate the (N)/year for Table 9.5.

2. If these numbers are significantly different, label the *y*-axis differently on either side of the ____ .

3. Plot the data in the graph **on the next page** and draw a line of best fit for each. Use different colors or dashed vs. solid lines to differentiate between the 2 data sets.

Use this graph for your Table 9.5 data.

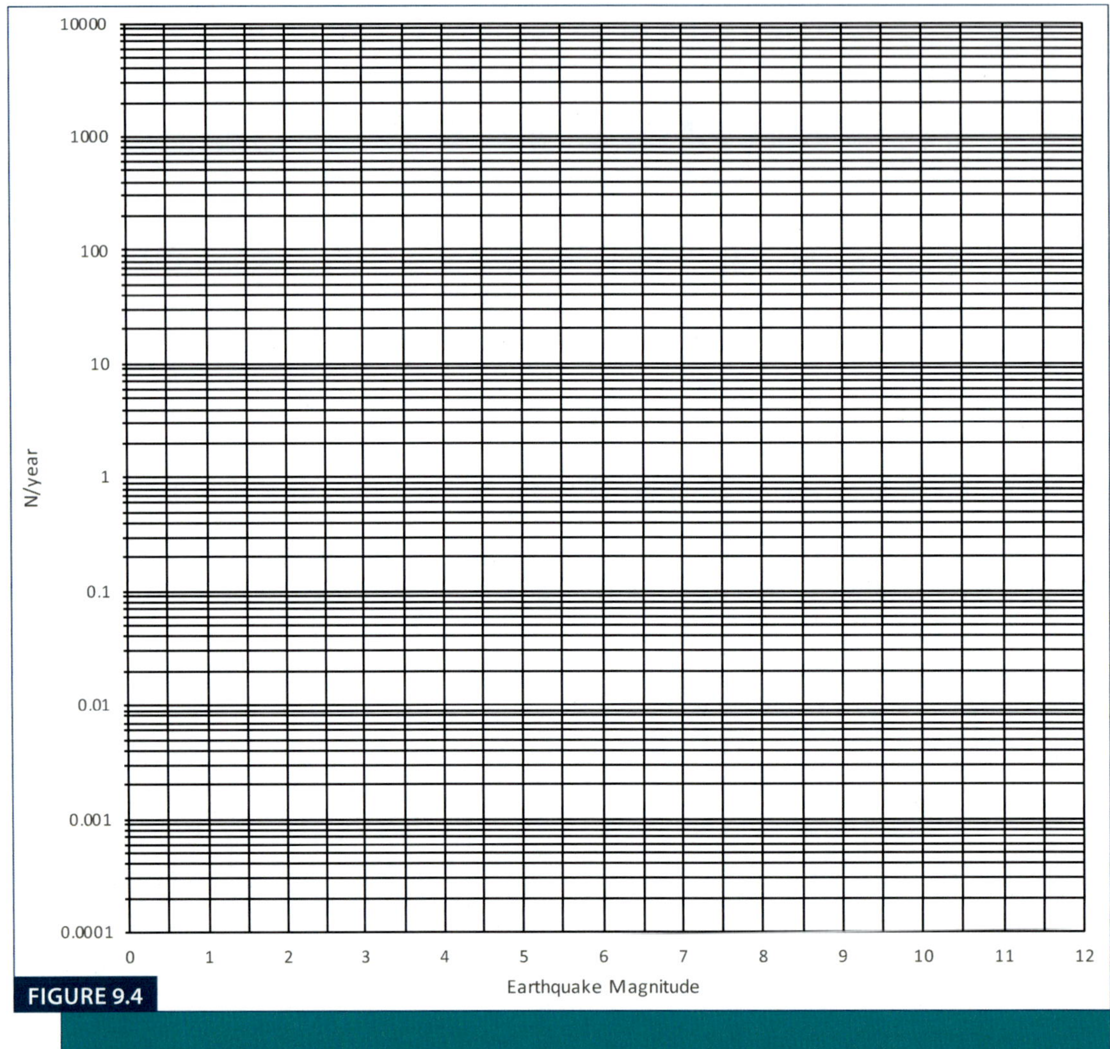

FIGURE 9.4

CONCLUSIONS

1. Compare and contrast the observed trends for the 2 regions:

   a. Which region has a higher probability of a magnitude 7 earthquake event?

      ⊙ Location 1 annual percent probability: _______________________________

      ⊙ Location 2 annual percent probability: _______________________________

   b. Compare the slope of the best fit line you drew for your Table 4 and Table 5 data sets. What does it mean if the slopes are the same? Different?

   c. Did your sampling technique influence your results on the graph?

      i. Was 30 years a long enough record to use for location 1? Why or why not?

      ii. Was 30 years a long enough record to use for location 2? Why or why not?

      iii. Was the size of the focus area for each region the same? If not, how does this impact the data? (Refer to the slope of the line of best fit and the (N)/year value.)

2. Does the elastic rebound earthquake model create a dataset that realistically replicates earthquake occurrence in the real world? Justify your answer and suggest other data that may have to be included in order to make the model more realistic.

10

GEOWORLD TECTONICS

Name: _________________________   Section: __________   Date: ________

Adapted from "Geoworld" Plate Tectonics Lab by Ann Bykerk-Kauffman

PREFACE

INTRODUCTION

Some of the **questions** asked in this lab are: **"Can we determine what Geoworld looked like in the past by identifying and age dating tectonic features?"** and **"How can we determine what Geoworld will look like in the future?"**

COURSE LEARNING OBJECTIVE ADDRESSED

Students will be able to …

- Analyze and interpret data pertaining to the study of Earth.
- Generate hypotheses regarding the geologic history of everyday landscapes.

SPECIFIC LEARNING OBJECTIVES FOR GEOWORLD: TECTONICS LAB

Students will be able to …

- Identify convergent, divergent and transform boundaries by identifying associated features visible on a map view.
- Calculate absolute rates of plate movement using absolute ages of hot spot volcanoes and calculate relative rates of plate movement using paleomagnetic data.
- Use calculated rates of plate movement to determine how geologic features have moved in the past and to predict future movement.
- Draw a cross-section (side view) of tectonic boundaries that illustrate the location of subducting plates and directions of movement within the mantle and lithosphere.

MATERIALS AND EQUATIONS

- This lab hand-out, printed
- Calculator (in the lab)
- Ruler (in the lab)
- To do conversions, remember km/Ma = mm/year

- Equation: rate = distance/time.
  - distance = rate × time, and
  - time = distance/rate

BEFORE YOU BEGIN

1. The diagram to the right shows a map view of a tectonic boundary.

   a. What type of tectonic boundary is shown here? (**Circle one.**)

      i. Divergent

      ii. Convergent

      iii. Transform

   b. Which feature is on the plate boundary? (**Circle one.**)

      i. Trench

      ii. Volcano

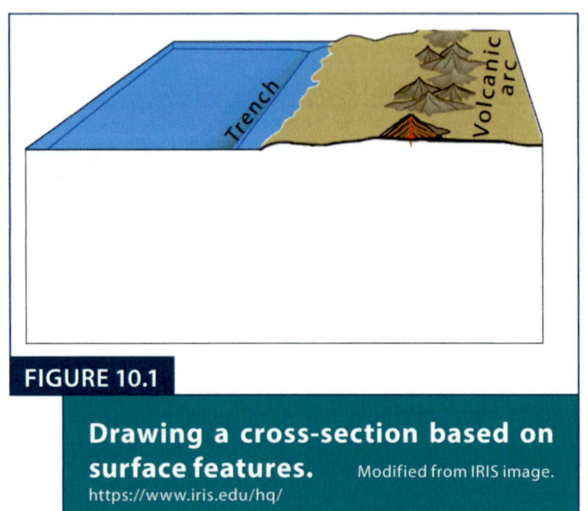

FIGURE 10.1

Drawing a cross-section based on surface features. Modified from IRIS image. https://www.iris.edu/hq/

   c. Using the pre-lab reading to help you, draw a cross-section (side view) of the boundary in the blank white panel at the front of the diagram.

2. The diagram on the right shows a map view of a divergent boundary that is offset by transform boundaries.

   a. There are 2 tectonic plates shown in this diagram. Trace out the boundary between the two plates.

   b. Sketch out how paleomagnetic signatures could look here, keeping in mind that the center of symmetry for paleomagnetic lines would be the divergent boundary.

   c. Draw arrows representing directions of relative plate movement along the transform boundaries.

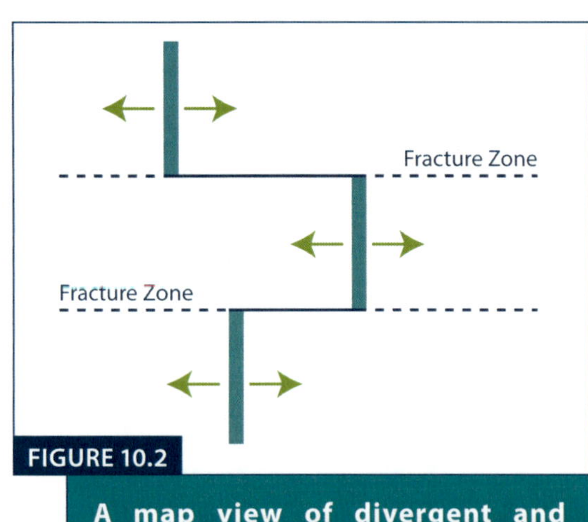

FIGURE 10.2

A map view of divergent and transform boundaries separating 2 tectonic plates.

3. The plate on the right side of diagram 2 has been growing at a rate of 2.5 km per million years (2.5 km/Ma).

    a. How fast is the ocean basin this ridge is in growing (in km/Ma)?

    b. Rearrange the rate = distance/time equation so that you are solving for time. Write it here:

        i.   The ocean is 700 km wide. Use the rate of growth (from 3a), and your rearranged equation from 3b. Determine when this ocean basin began forming. Show units.

    c. Rearrange the rate = distance/time equation so that you are solving for distance. Write it here:

        i.   A different ocean basin has a full spreading rate of 10 km/Ma. The oldest crust in this basin is 120 million years old. How wide is this ocean? Show units.

PART 1: PRESENT-DAY GEOWORLD

A map of Geoworld can be found in Figure 10.4. You may assume that the rates of plate motion on Geoworld (relative and absolute) are constant through time.

A. SEA FLOOR MAGNETIC ANOMALIES

Geoworld has a magnetic field aligned in the north-south direction. The polarity of this field reverses at random inter- vals through time, just as the one on Earth does.

The history of polarity of the magnetic field is well known for the past 150 mil- lion years and is shown by the adjacent magnetic polarity time scale. Shading represents normal polarity (the magnetic field points north) while white represents reverse polarity (the magnetic field points south). The sea floor on the map has also been shaded according to polarity of the rocks.

Note that the magnetic anomalies form a striped pattern on the map of Geoworld, and that the pattern in the Elrond Sea is symmetrical while that in the Aragorn Ocean is asymmetrical. This is because there is an active spreading ridge in the Elrond Sea but not in the Aragorn Ocean. The center of symmetry in the Elrond Sea is right along the spreading ridge.

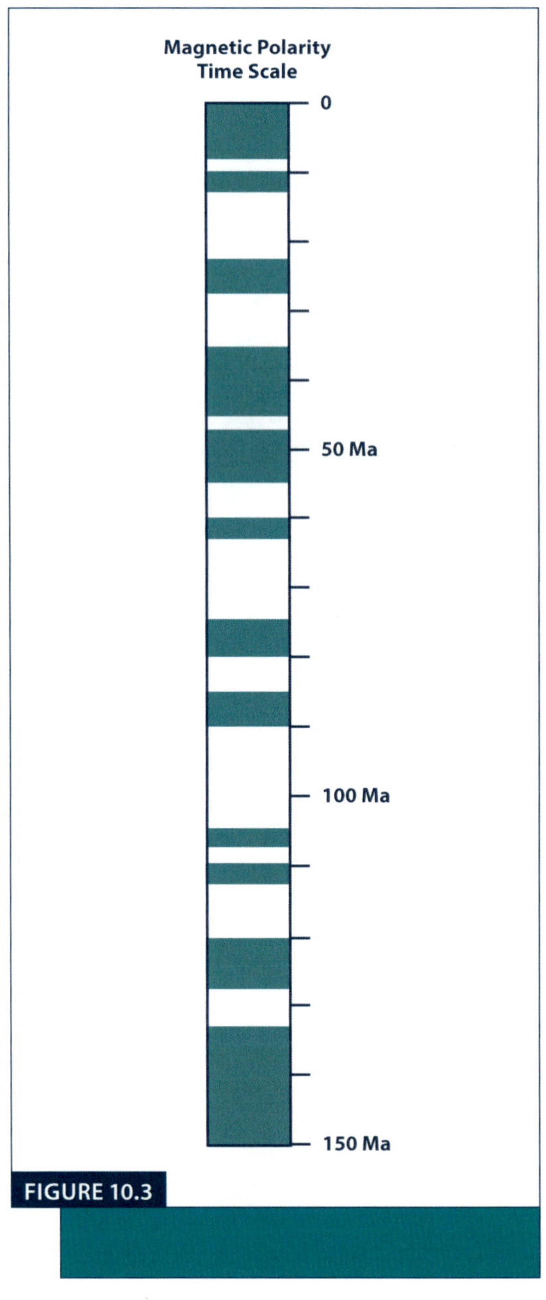

FIGURE 10.3

1. On the "Plate Tectonic Map of Geoworld," highlight the mid-ocean ridge in the Elrond Sea and label it the Hobbit Ridge. ***Hint:*** The middle of the mid-ocean ridge can be identified as it should be the center of symmetry. The mid-ocean ridge extends the length of the Elrond Sea, but is not a continuous line.

2. Is the rock at the center of Hobbit Ridge (older   /   younger   ) than the rock on either side of Hobbit Ridge? **(Circle one.)**

3. The pattern of positive and negative magnetic anomalies in the Elrond Sea can be correlated with the pattern of normal and reversed magnetism on the magnetic polarity time scale. Based on this correlation, how many million years (Ma) ago did sea floor spreading begin in the Elrond Sea?

   ANSWER: Spreading began _________________ Ma ago.

4. How fast is Frodo Continent moving away from Hobbit Ridge? This is the half spreading rate. We will solve this by breaking the problem into parts.

   a. Find the distance between the Frodo Continent and Hobbit Ridge on the map. _________________ cm

   b. Convert that map distance to a "real" distance on Geoworld using the map scale: 1 cm on the map = 400 km on Geoworld. Here's one way to set that up. Solve for x.

   $$\frac{1\text{ cm}_{(map)}}{400\text{ km}_{(geoworld)}} = \frac{(\text{measured \# of cm})_{(map)}}{X\text{ km}_{(geoworld)}}$$

   X = _________________

   c. How long has it been since Frodo Continent was sitting on Hobbit Ridge? **_Hint:_** This is your answer from #3. _________________

   d. Calculate the half spreading rate to determine how fast the Frodo Continent is moving away from Hobbit ridge. Use 4b as your distance and 4c as your time, solve for rate. Keep your answer in units of km/Ma.

   e. What is the full spreading rate of Hobbit Ridge (e.g., how fast are the Bilbo and Frodo Continents moving away from each other?) Keep your answer in units of km/Ma.

B. PLATE BOUNDARIES

5. There are two plate boundaries that act as **_major dividers_** of tectonic plates on the map. (_You already highlighted one of them—Part 1, #1._) Highlight the other major boundary.

6. Label both of the highlighted boundaries according to which type of boundary it is: divergent or convergent.

7. Highlight 3 active transform boundaries in the Elrond Sea (**_Only_** highlight where plates are actively moving in different directions), and label them with arrows indicating relative motion along the fault.

8. How many tectonic **plates** (not boundaries) are there total on Geoworld? _________

C. CROSS-SECTION

You've now identified plate boundaries based on surface features—but how do these look from a side view, inside the Earth?

9. Find A-A' on the map. Set another piece of paper on your map along that line. Mark where important features lie along that line. Then, move your paper to the area under the map where you're supposed to draw your cross-section. Use the marks on your paper to scale your drawing.

    a. If there are low elevation features along the a-a' line on your map, draw lower elevations in those spots on the cross-section (e.g., oceans).

    b. If there are high elevation features on the a-a' line on your map, draw higher elevations in those spots on the cross-section (e.g., mountains).

    c. If subduction is occurring, show the direction the subducting plate is travelling in the cross-section.

D. HOT SPOTS AND ABSOLUTE MOVEMENT RATES

The Gandalf Islands are analogous to the Hawaiian Islands on Earth. They form a linear chain that originated by the migration of a plate over a mantle hot spot. Mantle hot spots are assumed to be fixed relative to the center of the planet. All of the Gandalf Islands are volcanic in origin, but the easternmost and largest island—which is sitting directly over the hot spot—is the only one with active volcanoes on it. The rocks on the westernmost island have been dated using radiometric methods at 32 Ma, indicating that ***32 million years ago, that*** *island was sitting where the easternmost island is sitting today.*

10. **Calculate the absolute rate of plate movement (in km/Ma) of the Aragorn Ocean (and the Bilbo Continent, which is on the same plate).** Do this by measuring the distance between the centers of the westernmost and easternmost Gandalf Islands and comparing it to the age of the oldest island. Show your work. ***Hints:*** Do ***not*** use magnetic anomalies: use 32 Ma as the age of the westernmost island. Measure the distance on the map with a ruler, and convert map distances to actual distances using the same conversion as before:

$$\frac{1 \text{ cm }_{(map)}}{400 \text{ km }_{(Geoworld)}} = \frac{(\text{measured \# of cm})_{(map)}}{X \text{ km }_{(Geoworld)}}$$

11. From A.4.e. (page 116), we know that the Frodo Continent is moving eastward relative to the Bilbo Continent at a rate of ~40 km/Ma. You calculated the absolute rate of movement of the plate with the Bilbo Continent on it in question 10.

    a. Compare these absolute and relative rates of movement to answer this question: How quickly is the Frodo Continent moving relative to the Galadriel islands, and is the continent moving *towards* or *away* from them? Justify your reasoning.

PART 2: GEOWORLD IN THE PAST

Look for features on the map that may illustrate *previous* plate boundaries that are no longer active. Note that the Bilbo and Frodo Continents were combined to form the Bilbo/Frodo Supercontinent.

12. The Misty Mountains and the Rivendell Mountains are a non-volcanic mountain range with metamorphic rocks that formed ~200 million years ago.

    a. Are the two mountain ranges related, and if so, how?

    b. What type of plate tectonic event do they probably represent? ________________

13. While the Elrond Sea has been opening, the Aragorn Ocean has been closing. Assuming that (1) no sea-floor spreading has taken place in the Aragorn Ocean for at least 65 million years, and (2) subduction HAS been occurring for at least 65 million years, **how wide was the Aragorn Ocean along the A-A′ line at the time spreading started in the Elrond Sea?** Use the Sauron Trench as the western boundary of the Aragorn Ocean. *Note:* We will solve this by breaking the problem into parts.

    a. What is the current width of the Aragorn Ocean? Report your answer in km by converting your map measurement to a "real world" measurement:

$$\frac{1 \text{ cm }_{(map)}}{400 \text{ km }_{(Geoworld)}} = \frac{(\text{measured \# of cm})_{(map)}}{X \text{ km }_{(Geoworld)}}$$

   X = ________________

b. How much of the Aragorn Ocean has been subducted in the last 64 million years (since the Elrond Sea first opened)? **_Hint:_** You are given time, and you can use the absolute rate of motion you calculated in Part 1 D, #10.

E. THE FUTURE STATE OF GEOWORLD

14. When will the Bilbo Continent and the Sauron Trench first collide? Show your work. **_Hint:_** Use the absolute rate of motion you calculated in Part 1 D, #10 and the shortest distance between the Bilbo Continent and the Sauron Trench. You'll need to convert from map to "real world" measurements.

15. How wide will the Elrond Sea be at that time? Show your work. **_Hint:_** You will need to use the full spreading rate you calculated in Part A, #4 and the time you calculated in the previous question. The question is **_not_** asking how much wider it will become, but what the **_total_** width will be at that time.

FIGURE 10.4

A
A'
Galadriel
Islands
Sauron Trench
Rivendell Mountains
Elrond Sea
Misty Mountains
Aragorn Ocean
Equator
Frodo Continent
Bilbo Continent
Gandalf Islands
Gollum Island
N
Map Scale:
1 cm = 400 km
A
A'

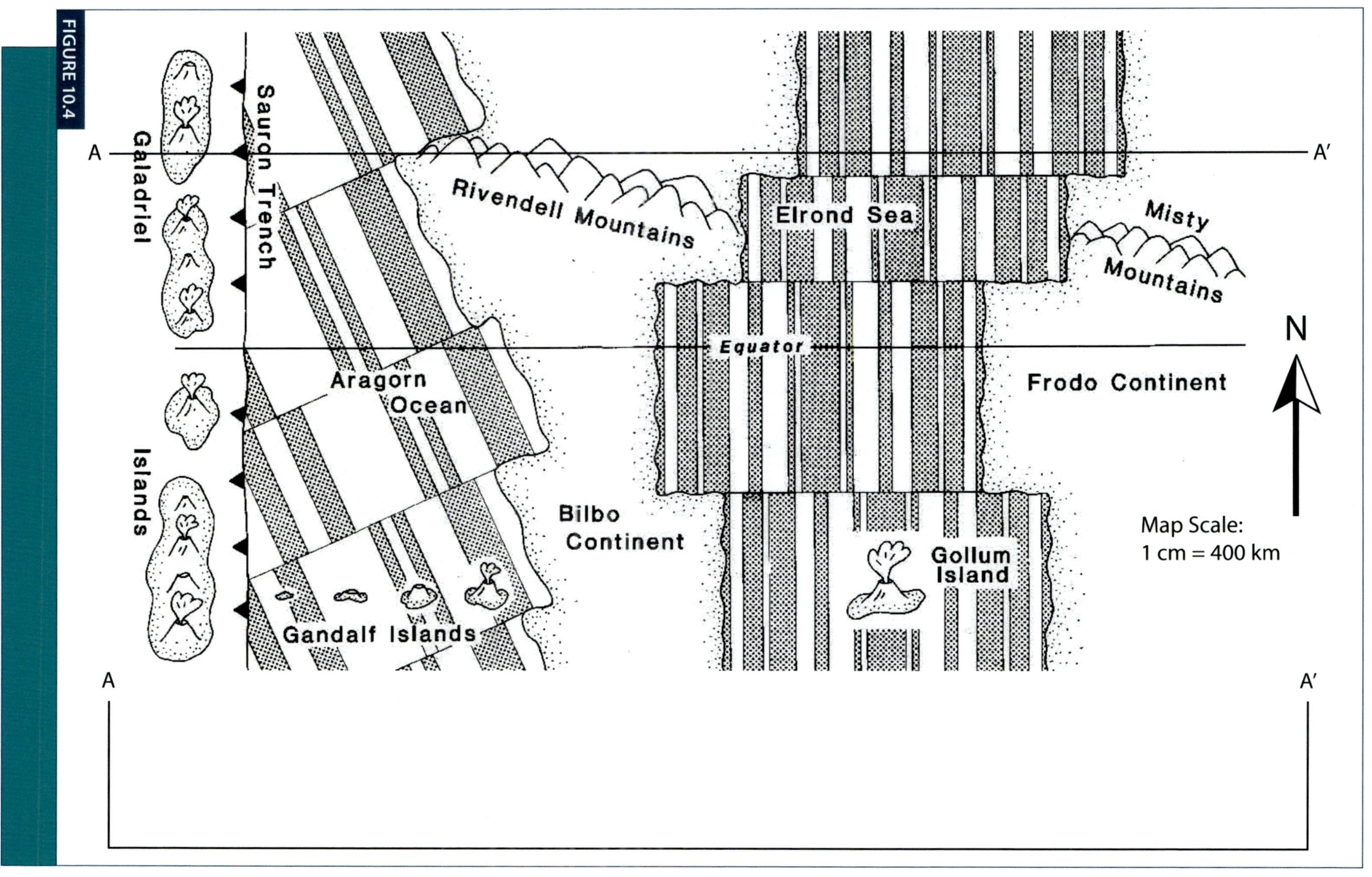